Introduction to Petroleum Production

Volume 2

Gulf Publishing Company • Book Division • Houston, London, Paris, Tokyo

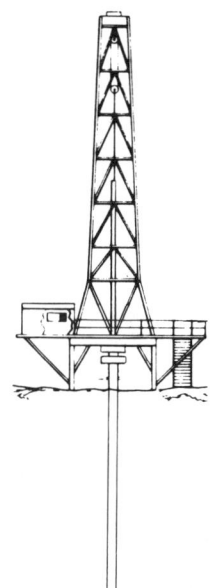

Introduction to Petroleum Production

Volume 2

Fluid Flow, Artificial Lift,
Gathering Systems,
and Processing

D. R. Skinner

Introduction to Petroleum Production

Volume 2
Fluid Flow, Artificial Lift,
Gathering Systems, and Processing

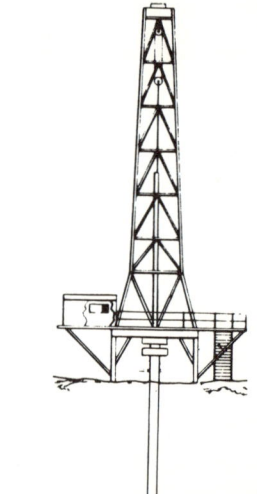

Copyright © 1982 by Gulf Publishing Company, Houston, Texas. All rights reserved. Printed in the United States of America. This book, or parts thereof, may not be reproduced in any form without permission of the publisher.

Library of Congress Cataloging in Publication Data (Revised)

Skinner, D. R., 1946-
   Introduction to petroleum production.
   Includes indexes.
                Contents: v. 1. Reservoir engineering, drilling, well completions—v. 2. Fluid flow, artificial lift, gathering systems, and processing.
   1. Petroleum engineering. I. Title.
            TN870.S533   622'.3382   81-6264
            ISBN 0-87201-767-2 (v. 1)   AACR2

Volume 1
Reservoir Engineering,
Drilling, Well Completions

Volume 2
Fluid Flow, Artificial Lift,
Gathering Systems, and Processing

Volume 3
Well Site Facilities: Water Handling,
Storage, Instrumentation,
and Control Systems

Contents

Chapter 1 Fluid Flow 1
    Reservoir Energy. Physical Properties of Reservoirs and Their Contents. Drilling and Completion. Wellbore Fluid Flow.

Chapter 2 Natural Flow 10
    Subsurface Production Equipment. Rate Control of Flowing Wells. Surface Equipment for Flowing Wells. Two-Phase Flow. Petroleum Recovery by Natural Flow.

Chapter 3 Artificial Lift 45
    Gas Lift. Plunger Lift. Sucker Rod Pumping. Tubing and Subsurface Equipment. Electric Submersible Pumping. Hydraulic Pumping Systems. Artificial Lift System Analysis. Effect of Natural Flow on Artificial Lift. Casing Vacuum Pumps. Comparison of Artificial Lift Systems. Fluid Transportation.

Chapter 4 Surface Gathering Systems 114
    Types of Gathering Systems. Fluid Flow Behavior. Flowlines. Headers. Valves. Flow Behavior in Gathering Systems. Gathering System Design.

Chapter 5 Gas Processing 139
  Two-Phase Separators. Dehydrators. Gas Compressors. Prime Movers. Gas Sweetening Equipment. Sulfur Recovery Units. Condensate Separation. Vapor Recovery from Storage Tanks. Gas Process Facilities.

Chapter 6 Liquid Processing 175
  Oil-Water Emulsions. Free-Water Knockout. Treaters. Safety Precautions With Treaters. Other Treating Methods. Oil and Water Handling.

Chapter 7 Liquid Storage 202
  Liquid Storage Tanks. Liquid Pumps. Small-Lease Production Facilities. Large Production Facilities.

  Index 228

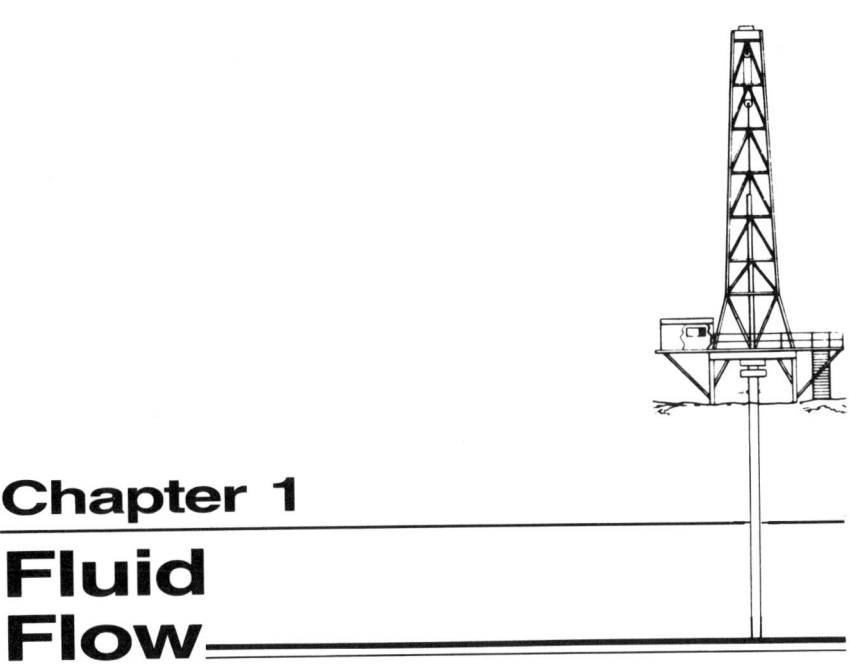

Chapter 1
Fluid Flow

In Volume 1 of this series we studied the nature of oil and gas, how they are created and stored in reservoirs, and how those reservoirs are found and accessed. We also looked, briefly, at the ways in which the fluids are brought to the surface.

In this second volume we will take a detailed look at the principles and technology involved in getting the oil and gas out of the ground, to the surface, and ready for transportation to the refinery or gas processing plant. As an introduction to all this, it will be helpful to quickly review the physical and mechanical properties of reservoirs. Much of this information appears in Volume 1, with more illustration and detail.

Reservoir Energy

Petroleum was formed when ancient plant and animal remains were submerged and buried beneath prehistoric rivers and seas, compressed by millions of tons of rock, and distilled by millions of years of exposure to high temperature and pressure. The result of this distillation process is a mixture of hydrogen and carbon molecules (hydrocarbons). The mixture of hydrocarbons can take the form of natural gas or liquid crude oil. Water that was mixed with the original organic remains or that moved into the petroleum-bearing reservoir can be mixed with hydrocarbon gases and liquids.

Energy is stored in the fluids (gases and liquids) and rock in the reservoir as a result of the pressure of the overlying layers of earth and the heat of the earth's core. This energy is manifested as high pressure and temperature in the reservoir. The reservoir pressure represents the driving force which causes fluids to flow from the heart of the reservoir to a well. Without reservoir pressure, there can be little or no recovery of petroleum.

As long as the reservoir pressure is high enough, oil and gas are pushed to a wellbore from which they can be recovered. Removing petroleum from a reservoir when the only energy source is the initial reservoir energy is called *primary recovery*. Figure 1–1 illustrates fluid flow in primary recovery.

As fluids are removed from the reservoir, the pressure decreases. Petroleum which remains after the reservoir pressure is too low to push fluids to the well must be recovered by transmitting energy to the reservoir from man-made sources. This is called *secondary recovery*.

The first methods of secondary recovery utilized gas injection and a primitive form of water injection—saltwater disposal. Later, water injection or secondary waterflooding became the dominant form of secondary recovery (see Volume 1, pp. 32–35). Recently, a number of improved recovery methods have been developed which include injection of viscous polymers and mixtures of gases to form miscible mixtures of oil, gas, and water. The techniques used for recovery—even water injection—have been developed to such a level that most induced recovery techniques are now referred to by the term *enhanced recovery*.

Figure 1–2 illustrates the flow of fluids when secondary or enhanced recovery methods are used. Fluids such as water, gas mixtures, and chemical compounds are injected into the reservoir to form fluid banks which push reservoir fluids to nearby production wells. In every type of enhanced recovery, energy is supplied from a man-made source to the reservoir.

Physical Properties of Reservoirs and Their Contents

When a petroleum reservoir is discovered, the native properties of the reservoir may not allow petroleum to be removed easily. It is not realistically possible to change some of these natural properties, such as saturations, wetting tendencies, and relative permeabilities.* However, it is possible to change some properties—particularly permeability and to some extent porosity—in the critical flow area near the wellbore, thus increasing the flow capacity of the well. Another way to increase flow capacity is to extend the surface area through which fluid flows into the well.

---

*For a discussion of these properties, see Volume 1, pp. 16–20.

Fluid Flow 3

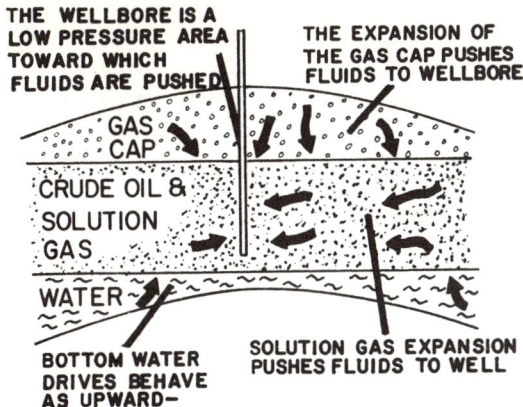

Figure 1-1. Original reservoir energy pushes fluid to wells in primary recovery.

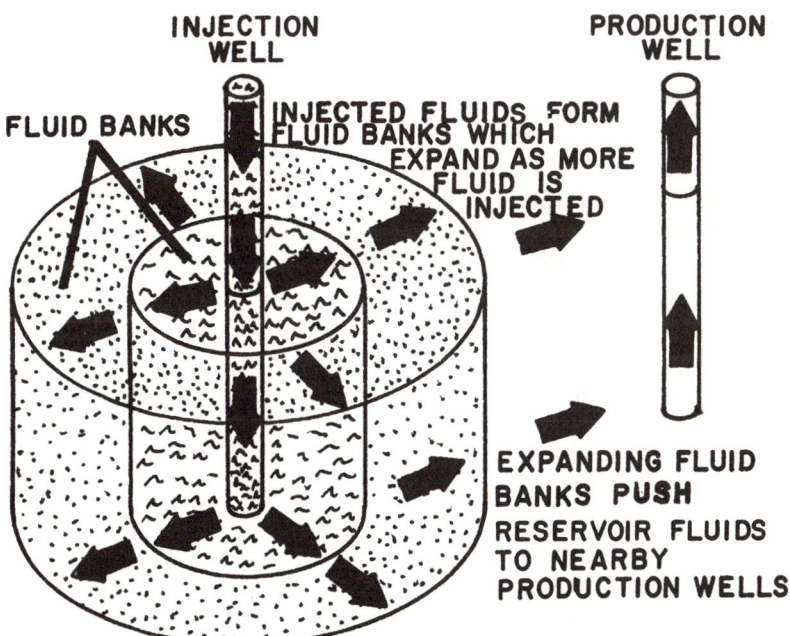

Figure 1-2. Secondary and enhanced recovery methods utilize radially expanding fluid banks to push reservoir fluids from injection wells to producing wells.

4   Introduction to Petroleum Production

The flow capacity of a well may be increased by stimulating the well at the time it is drilled or at any time in its productive life. There are several methods and combinations of techniques that may be used to stimulate a well. The most notable are acidizing, hydraulic fracturing, and explosive fracturing. (See Volume 1, Chapter 5).

Selective Stimulations

Petroleum reservoirs are often composed of layers of productive intervals separated by non-productive or impermeable layers. The need for stimulation may not be the same for each interval. Selective stimulations are used to treat intervals of the reservoir separately.

Selectively stimulating a well usually consists of slightly treating the most productive or permeable layers, temporarily blocking these layers, treating less permeable layers, blocking them, and so forth. Some of the methods for blocking intervals are: using ball sealers (rubber balls which plug perforations much like corking a bottle), using packers and bridge plugs, using sand plugs, using temporary blocking gels, and using interface pumping techniques.

Selective stimulation can be used with acid stimulations and with fracturing techniques. The effectiveness of blocking methods depends on the stimulation technique used, the rate and pressure used, and the wellbore or casing conditions.

Flow-Induced Properties

As petroleum and water are removed from a reservoir, some of the reservoir and fluid characteristics change. For example, the saturations in parts of the reservoir change as fluids move. As the saturations change, the relative permeabilities also change.

As fluids are removed from a reservoir, the pressure usually decreases. The density, viscosity, and surface tension of hydrocarbons are affected by pressure changes. Thus, as fluids flow from a reservoir, the fluid as well as reservoir characteristics change.

Using a complex mathematical derivation, it can be found that some 80% of the pressure change from the reservoir pressure to the wellbore pressure occurs within a few feet of the well. Anytime the pressure of a compressible fluid like gas or gas-saturated oil decreases, its temperature also decreases. In the few feet of reservoir near the wellbore there can be a temperature differential of 10–100 degrees. This temperature differential can cause precipitation of solid scale compounds (salts of calcium, sodium, magnesium, and barium) which are normally dissolved in reservoir liquids, and the solidification of paraffin—heavy, asphaltic hydrocarbons—in the pore space of

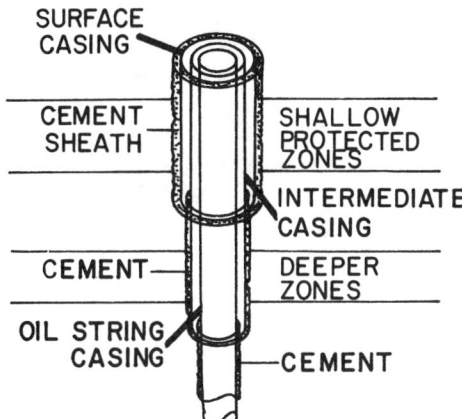

Figure 1-3. Several casing strings are usually required to protect the wellbore and surrounding formations from interference.

the reservoir. These depositions reduce or completely block the permeability near the wellbore and reduce or stop fluid flow altogether. Unfortunately, stopping flow temporarily allows the temperature to increase but does not remove the deposits; stimulation or chemical treatment is usually required to restore acceptable flow characteristics.

Drilling and Completion

State and/or federal regulations usually govern the methods used to drill a well while isolating formations, such as fresh-water aquifers, through which the wellbore passes. These formations are blocked using one or more casing strings which are installed and cemented in place before deeper drilling proceeds.

The surface casing is used to isolate shallow, fresh-water aquifers. Intermediate casing is used to isolate deeper formations which interfere with later drilling operations. The casing string which extends to the petroleum reservoir is called the oil string casing. Figure 1-3 shows the way in which these casing strings are installed.

When the potentially productive zone(s) of a formation have been reached by drilling, the oil string casing is set in place with cement. Depending on the well completion method used, this casing may be perforated, and, if necessary, sand control technology may be employed. (See Volume 1, pp. 69–95.)

Wellhead Arrangement

All casing strings and any tubing in a well must be connected at the earth's surface. The equipment used to connect casings and tubing and to seal the hydrocarbons inside them is called the *wellhead* (see Figure 1–4).

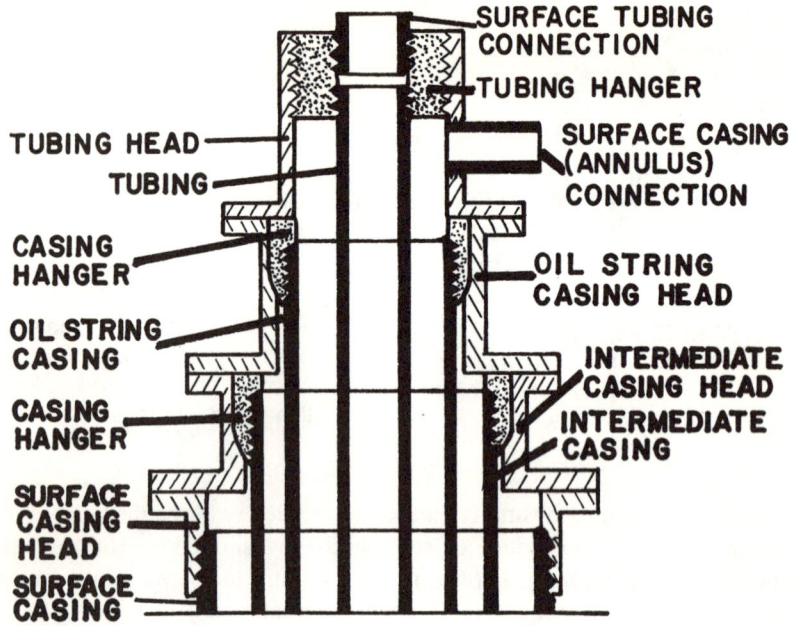

Figure 1-4. A wellhead is used to connect all casing and tubing strings in a well.

Wellbore Fluid Flow

Hydrostatic Pressure of Liquid Columns

Liquid exerts a pressure that varies with depth. The pressure exerted by liquid is greater at the bottom of the liquid than at shallow levels. The pressure exerted by a liquid depends on the type of liquid and the depth. The pressure exerted by a column of liquid and its weight is expressed by several parameters. These parameters are all related and are simply several methods of giving the same information.

Density. The density of any substance is the weight of a given volume of the substance. Density is expressed in pounds per cubic foot (in the English system) at a standard temperature of 60° F. The density of pure water is 62.4 pounds per cubic foot.

Specific Gravity. Specific gravity is the ratio of the density of a substance to that of plain water. For example, the specific gravity of crude oil is about

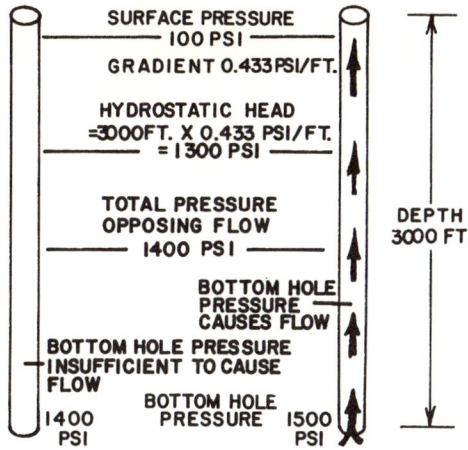

Figure 1-5. Fluid flows into a wellbore if the reservoir pressure is greater than the pressure at the bottom of the wellbore.

0.85 because its density is about 53.04 pounds per cubic foot compared to the density of water—62.4 pounds per cubic foot (53.04 = 85% of 62.4). The specific gravity of pure water is 1.0. The specific gravity of water completely saturated with salt is about 1.4.

Gradient. The gradient of a liquid is the pressure exerted by a column of liquid one foot deep. The gradient depends only on the density of the liquid, not the size of the container. The gradient of any liquid is the product of the liquid's specific gravity and 0.433 psi per foot.

Hydrostatic Head. A column of liquid exerts pressure at a depth in the column that depends on the liquid's density and the depth. The pressure at the bottom of a column of liquid is called the hydrostatic head of the liquid. The hydrostatic head of a column of liquid is the product of the liquid's gradient and the depth of the column. The hydrostatic head of a 1000-foot-deep column of pure water is 434 psi, while the hydrostatic head of the same column filled with crude oil is 368 psi.

Fluid Flow into a Wellbore

Fluid flows to a well because the reservoir pressure is higher than the pressure in the well. Liquid flows upward if the pressure below the liquid exceeds the sum of the pressure above the liquid and the hydrostatic head of the liquid. This is illustrated in Figure 1–5.

Types of Fluid Flow

When fluid moves upward slowly and simply (illustrated in Figure 1–5), the flow is called hydraulic flow or simply liquid flow. Liquid seldom flows alone in an oil well; gas is also involved in the fluid flow. There are several types of flow involving gas. These can occur not only in a wellbore but also in surface equipment, and the type of flow affects equipment performance.

Aerated Flow. When free gas is mixed with and flows with liquid, the flow is said to be aerated. This flow is liquid containing a great number of bubbles.

Slugging Flow. Occasionally, gas and liquids will flow alternately in a well. Slugs of liquid will be followed periodically by slugs of gas. This type of flow can cause damage to improperly designed or installed equipment because the slugs of liquid can strike equipment with the force of a sledge hammer in a phenomenon called *water hammer*. Slugging flow is characterized by fairly short intervals of liquid and gas flow.

Heading Flow. Heading flow is similar to the cyclic pattern of slugging flow, except that the cycles may be as long as several hours. Heading flow has an adverse affect on subsurface and surface equipment when equipment designed for liquid service has only gas entering for several hours.

Blowing Flow. In this case a large volume of gas carries only a spray of liquid much like the discharge of an aerosol can.

Reservoir Flow

The flow pattern in a wellbore (and eventually in surface equipment) is caused by flow patterns in the reservoir. Aerated and blowing flow result when gas and liquid mix in the reservoir and as they enter the wellbore. Cyclic flow patterns such as slugging and blowing occur when a periodic rhythm of wellbore hydrostatic pressure and reservoir pressure develop. This cyclic pressure pattern usually affects the reservoir by setting up wave-like action much like the oscillatory motion of a waterbed.

Fluid Flow to the Surface

For fluid to flow from the bottom of a well to the surface, the bottom hole pressure must exceed the sum of the surface pressure and the hydrostatic head of the fluid column. If the bottom hole pressure is high enough, the liquids will flow to the surface. This type of recovery is called *natural lift* or *natural flow*. If there is not enough bottom hole pressure for natural flow, the liquids must be lifted from the well using man-made energy sources. This method of liquid recovery is called *artificial lift* or *induced flow*. Natural flow is the subject of the next chapter.

Chapter 2
Natural Flow

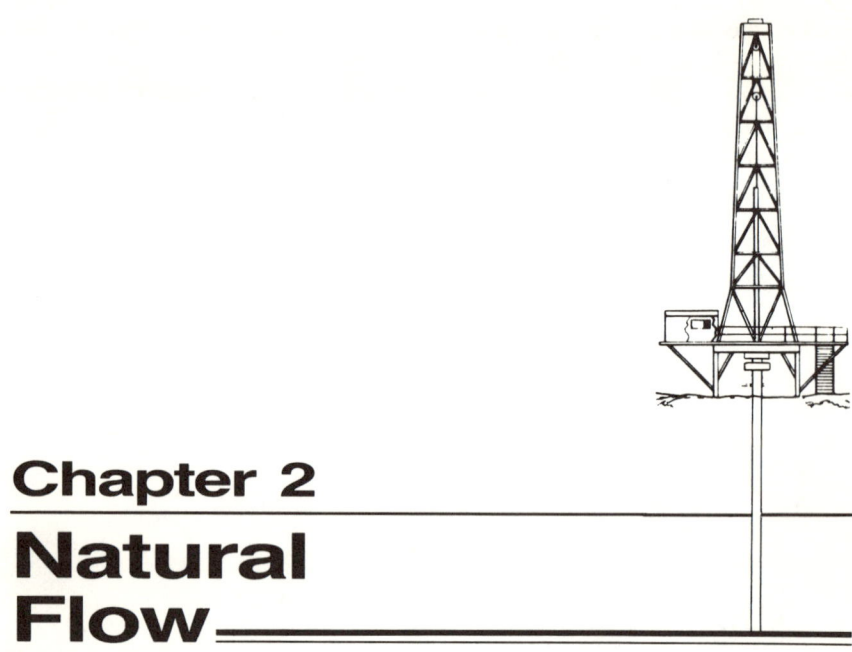

Petroleum can be recovered from a well by natural flow only when bottom hole pressure exceeds the hydrostatic head of the fluids in the well plus the surface pressure. When the flow of liquids stops, there may still be considerable bottom hole pressure because the hydrostatic head of several thousand feet of liquid is more than 1000 psi.

Natural flow is the only production method used for reservoirs containing gas only. When the natural flow of gas stops, the hydrostatic head of the gas plus the surface pressure exceeds the low reservoir pressure. The gradient of natural gas is about 0.00056 psi per foot, and the hydrostatic head of gas in a deep well (20,000 feet in depth) would be only 11.2 psi. For comparison, sea level atmospheric pressure is about 14.7 psi. When gas stops flowing naturally, there is not enough gas left to attempt other recovery methods.

Subsurface Production Equipment

Figure 2–1 is a typical configuration of subsurface equipment in a naturally flowing production well commonly called a *flowing well*. The major items of production equipment are the tubing string and packers or bridge plugs. Although the size and condition of the oil string casing must be known to use other equipment, the casing is not normally thought of as part of the production equipment (see Volume 1, Chapter 6).

Natural Flow 11

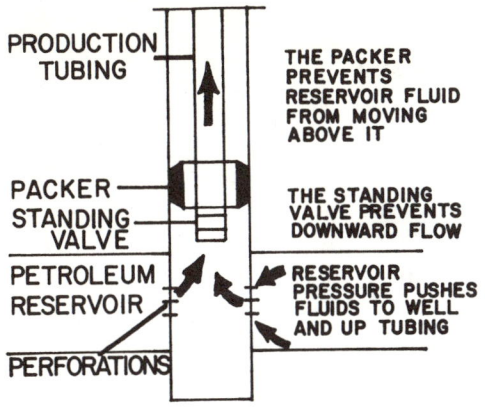

Figure 2-1. A naturally flowing well is usually equipped with tubing to conduct fluid and a packer to isolate the upper casing from high bottom-hole pressure.

Tubing

The pipe used for production tubing is not usually the same as the standard pipe one sees in normal surface use. Because of the conditions and loading of production tubing, a special pipe design is required. A joint of production tubing must be able to bear the weight of all tubing below it. Tubing must be able to withstand high internal pressure of several thousand psi without bursting and be able to withstand external pressure in the same range without collapsing.

Metallurgy. Minimum yield is one factor used in judging the strength of tubing. Steel can be stretched or compressed when very heavy forces are used; however, when the force is removed, the steel should return to its original shape. Yield is the maximum force per unit area to which steel can be subjected and return to its original shape. Exceeding the yield point of steel means it will be permanently deformed. Yield is measured in pounds per square inch (psi), but these units do not have the same meaning when applied to pressure.

In Figure 2-2 a piece of tubing with an outside diameter (OD) of 2.5 inches and an inside diameter (ID) of 2.0 inches is supporting a load of 50,000 pounds. The cross-sectional area of the steel is 1.77 square inches, and the load on the steel is 28,294 pounds per square inch. If the yield of the steel were 30,000 psi, the steel would stretch slightly, like a rubber band; when the load is removed, the steel returns to its original shape. On the other hand, if the yield were 20,000 psi, the steel would be permanently deformed by this load.

The yield of steel depends on the metals and the methods used to make the steel alloy. The yield as well as properties affecting use in production opera-

12 Introduction to Petroleum Production

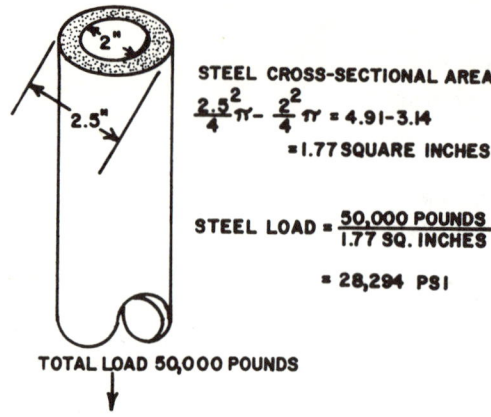

Figure 2-2. The cross-sectional area of tubing must be able to support the weight of all tubing below it without breaking.

tions, such as hardness, tensile strength, and resistance to corrosion, are used to determine the grade of tubing.

The American Petroleum Institute (API) has established a set of standard tubing grades used throughout the industry. Table 2–1 shows the most commonly used grades of tubing. The number in the grade is the minimum yield of the steel in thousand psi. The yield of a particular joint of tubing may be more (but never less) than that specified in the grade, thus the name "minimum yield strength."

Mechanical Configuration. The individual joints of tubing must be linked together to make a tubing string. The coupling method used depends on the weight to be borne by the tubing, the outside diameter of the tubing and coupling (collar) and the space into which the tubing must fit, and the inside diameter required.

When the threads are placed in the tubing, the threads reduce some of the cross-sectional area designed to support the weight on the tubing. Without any special provisions, the part of the tubing joint into which the threads were cut becomes the weakest part of a joint.

In most cases it is not advisable to cut the threads and create a weak part of the tubing. It is possible to add additional wall thickness to the tubing in the area where threads are to be cut. Thus, when threads are in place, the thread area of a tubing joint has as much or more cross-sectional area and is as strong as any other part of the joint. This thickened part of the tubing is called an *upset*.

The additional wall thickness may be added to the outside of the tubing, in which case the tubing is called *externally upset tubing* and has the standard designation EUE. The upset of EUE tubing has the same inside diameter as

Table 2–1
Commonly Used Grades of Tubing

Tubing Grade	Internal Yield Strength (psig)
F–25*	25,000
H–40	40,000
J–55	55,000
C–75	75,000
N–80	80,000
P–105*	105,000

*Not API standard
(Courtesy of the American Petroleum Institute)

the tubing body, and any equipment that can pass through the body of the tubing can also pass through the upset.

The upset could be placed on the inside of the tubing, but the ID of the upset is less than that of the body and acts as a restriction in the tubing. Internally upset tubing is not used for normal production activity. Figure 2–3 shows the appearance of EUE, internally upset, and non-upset (NU) tubing.

The threaded part of the tubing joint, called a male thread, is mated with the female thread of a tubing collar in standard tubing installations. The tubing threads are tapered slightly so that when the tubing and collar are tightened, the collar expands and the male thread compresses slightly. This slight stretch causes the tubing thread and the collar to be in tension constantly. The threads are designed so that this force makes the best possible pressure seal and minimizes the possibility of the tubing and collar unscrewing due to vibration. Figure 2–4 illustrates the use of tapered threads.

There are a number of thread styles available for tubing and most other types of pipe. The most common types of tubing threads are 8 RT and 10RT, meaning 8 and 10 round threads per inch. Another type of thread occasionally used for tubing is called a *butress thread*. Because of exposure to dirt and other contaminants during installation and removal, round threads are often used, since they are less susceptible to contamination than other thread types.

Tubing is sometimes installed in locations where there is very little available space, and the OD of the collar may be too large to fit. Other coupling methods are available for such situations. One such method is to use thin-wall collars, but this is usually the least satisfactory, since only a fraction of the collar may be removed and still have a strong coupling. Figure 2–5 shows flush joint tubing, a special type of tubing used for slim hole applications. This tubing has specially made male threads and collars. Figure 2–6 shows the use of integral joint tubing, a type of tubing that does not use collars.

14   Introduction to Petroleum Production

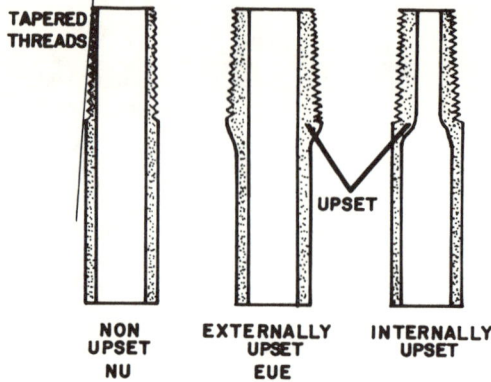

Figure 2-3. Several thread styles are used with tubing in wells.

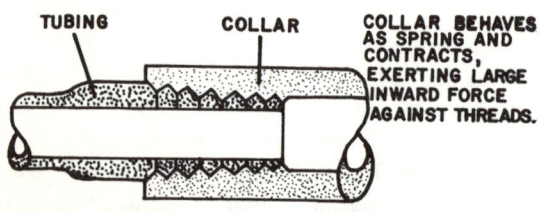

Figure 2-4. Tapered threads are used to form strong, tight seals between tubing joints.

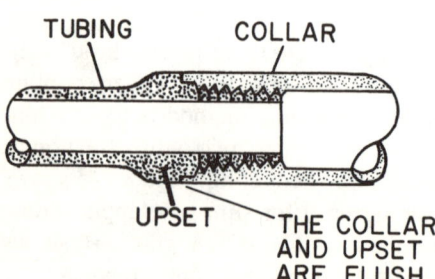

Figure 2-5. Flush joint tubing is used when there is too little clearance between the casing and tubing to use upset joints.

Natural Flow 15

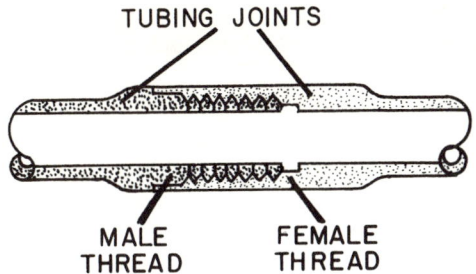

Figure 2-6. Integral joint tubing is sometimes used when there is little clearance in a wellbore.

The types of tubing used for slim hole operations are usually expensive, and they are significantly weaker in the coupling area than in the tubing body. If this weakness is considered in the tubing installation and the appropiate precautions taken, slim hole tubing can be safely used.

Dimensions and Strengths. The ratings for steel tubing are based on the tubing in tension, that is, any forces that are working to stretch the tubing. Although tubing may be installed in compression, the ratings of the steel are reduced significantly.

Tubing may be obtained in virtually any imaginable size. There are, however, some sizes that are commonly used. Table 2–2 shows some of the often-used tubing sizes and grades, along with the dimensions, strengths, and pressure ratings.

Production Packers

Production packers are devices which are attached to tubing and are used to isolate fluid streams in a well. Most production packers are constructed to be used in casing, but a number of open hole packers are also available.

Figure 2–7 is a mechanical production packer made for operation in casing. When this packer is at the desired depth, the upper slips are just touching the casing. The tubing is then rotated clockwise about 90° and raised about six inches. This action engages a set of steel teeth called *slips* (which catch the inside of the casing much like the jaws of a pipewrench grip a pipe) and pulls the metal section into the packing elements, inflating them. Tension is then applied to the tubing, and the tubing head is installed. This action leaves the tubing pulled taut.

The packer shown in Figure 2–7 has two sets of slips, one to be used when the packer is set in tension and the other to be used when the packer is set in compression. Although it is not a normal practice, it is sometimes necessary to set the packer with the tubing in compression. This is done when high pressure in the annulus opposes the force holding the packer in place.

Table 2–2
Dimensions and Strengths of Commonly Used Tubing

Nominal Tubing Size (inches)	API Grade	Outside Diameter (inches)	Inside Diameter (inches)	Collapse Pressure (psig)	Internal Burst (psig)
2.0	H–40	2.375	1.995	4160	3500
2.0	J–55	2.375	1.901	7180	7700
2.0	N–80	2.375	1.773	12,890	14,970
2.5	H–40	2.875	2.347	5230	5280
2.5	J–55	2.875	2.347	6800	7260
2.5	N–80	2.875	2.347	9420	10,570
3.5	H–40	3.500	2.992	5050	5080
3.5	J–55	3.500	2.992	6560	6980
3.5	N–80	3.500	2.992	9080	10,160
4.0	H–40	4.000	3.476	3220	2870
4.0	J–55	4.000	3.476	4420	4580
4.0	N–80	4.000	3.476	7780	9170
4.5	H–40	4.500	3.958	3930	4220
4.5	J–55	4.500	3.958	5100	5800
4.5	N–80	4.500	3.958	6810	8430

(From Halliburton Cementing Tables, Halliburton Services)

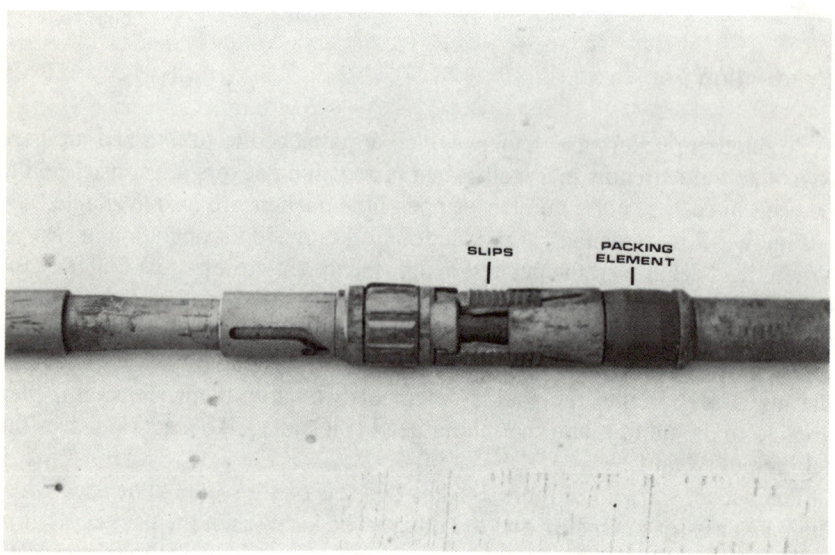

Figure 2-7. A mechanical packer prevents fluids in the tubing-casing annulus above the packer from communicating with those below.

Natural Flow   17

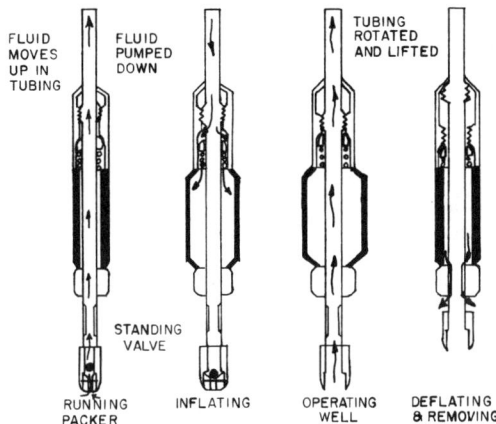

Figure 2-8. An inflatable packer is one type of packer used to isolate the tubing from the annulus.

Figure 2–8 is a mechanical view of an inflatable packer being lowered into a well, inflated, deflated, and removed from a well. Such a packer can be used in cased or open holes. This packer has no slips and its packing elements are inflated by wellbore liquids.

When packers are inflated in a wellbore, the pressures above and below the packer can exert upward or downward force against the packer. Figure 2–9 shows how such a difference in pressures can exert a force against the packer. If this force is great enough, the packing elements and the slips can lose their grip on the casing, and the fluid seal of the packer is lost. This is called "unseating the packer." A packer is set in tension or compression (depending on the direction of forces) to overcome the tendency to unseat it.

Bridge Plugs

It is occasionally necessary to block the bottom part of a cased hole permanently. Figure 2–10 shows a bridge plug, a device which uses slips and packing elements like those of a packer to divide fluid levels in a wellbore. Unlike a packer, once a bridge plug is set (the packers and slips engaged), the tubing is removed, and the seal is left in the wellbore permanently.

Tubing Anchors

Tubing anchors have slips like those on packers but have no packing elements. They are used to connect the bottom of the tubing to the casing via the anchor so that the tubing may be left in tension—even when a seal is not required in the annulus. Tubing anchors do not restrict flow in the annulus.

18 Introduction to Petroleum Production

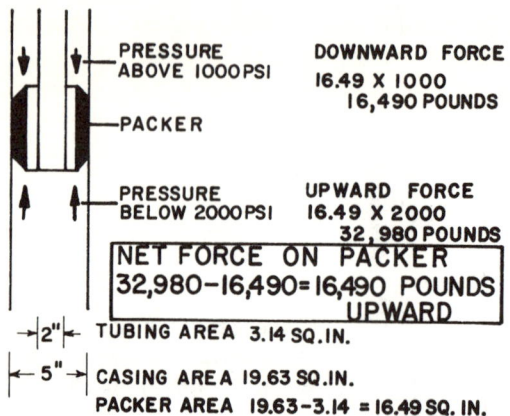

Figure 2-9. A difference in pressures above and below a packer can exert considerable force against a packer.

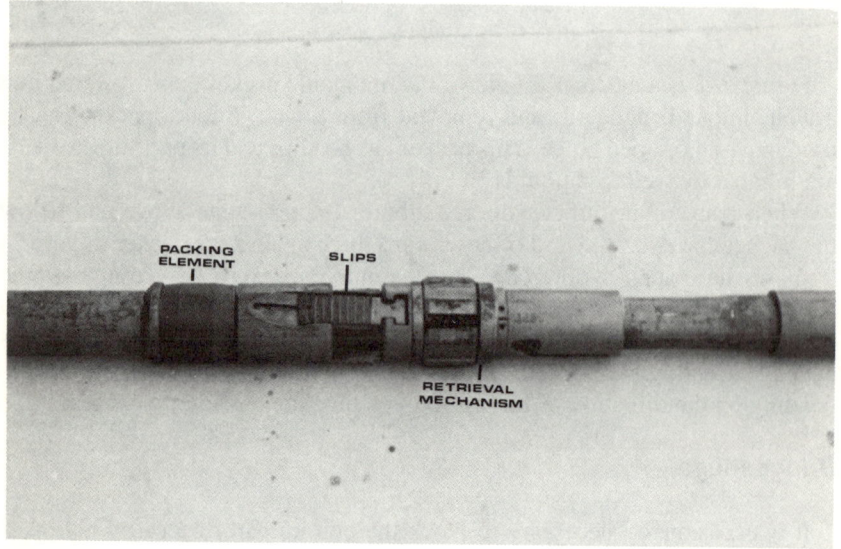

Figure 2-10. A bridge plug is similar to a mechanical packer but is left in the wellbore to block flow of fluids around it permanently.

Although anchors are sometimes used in flowing wells, they are usually installed in wells using artificial lift.

Seating Nipples

Seating nipples are special joints of tubing with significantly reduced inside diameter. Some tools used in tubing would fall through the bottom of the tubing string if no provisions were made to stop them. Seating nipples are used as the point at which subsurface equipment is to stop. Near the bottom of equipment to be used with a seating nipple is a slightly enlarged section called a *no-go*. The no-go will not pass through a seating nipple—even if the remainder of the equipment might. Figure 2–11 is a sketch of a seating nipple.

Standing Valves

When fluid enters the tubing of a flowing well, it must be able to move into and up the tubing without restriction. However, if the bottom hole pressure decreased or if the pressure in the tubing increased, the fluid would have a tendency to flow downward into the reservoir. If such an undesirable condition could possibly occur, it would be advisable to equip the well so that upward flow in the tubing is allowed while downward flow is prevented.

A standing valve is a one-way valve used to allow upward but not downward flow. The valve consists of a seat, a ball, and a cage as shown in Figure 2–12. The name "standing" comes from the fact that the valve is installed in the bottom of the tubing string and does not move. Figure 2–13 illustrates the operation of a standing valve.

A complete standing-valve assembly consisting of the seat, ball, and cage is installed as a unit. The standing valve is then mounted in the seating nipple, before or after the tubing is in place, and all fluid that enters the tubing must pass through the valve. Upward movement of fluid is not impeded by a standing valve, but downward flow is prevented.

Standing valves are sometimes dropped into tubing from the surface for the purpose of testing the tubing for leaks. After the standing valve has had time to fall to the seating nipple, liquid is pumped downward into the tubing against the standing valve. If the tubing is not leaking, the standing valve prevents flow, and the tubing pressure rises and is trapped when the pump is stopped. If the pressure drops, the tubing is known to be leaking. When the pressure test is complete, the standing valve may be recovered by lowering a grapple-like fishing tool, catching the fishing head on the valve, and removing the valve and tool.

Screens and Slotted Liners

Standing valves, other subsurface equipment, and some surface equipment are designed to operate with precise metal-to-metal seals. Any solid material can completely negate the operation of the equipment. Solids, such as sand or formation particles, can enter the tubing. Screens and slotted liners may be attached to the bottom of the tubing string to prevent the entry of solids into tubing. These devices act as filters in the well.

20 Introduction to Petroleum Production

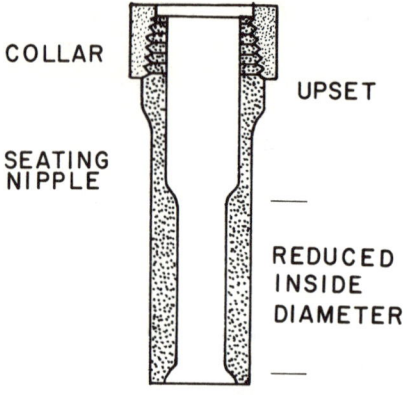

Figure 2-11. A seating nipple is attached to tubing to hold equipment used in tubing.

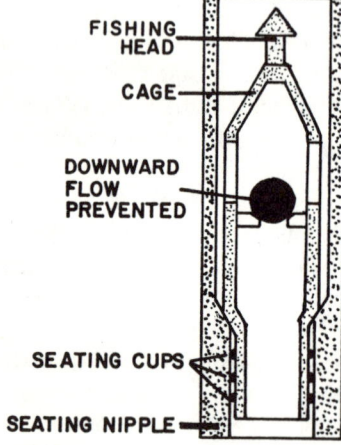

Figure 2-12. A standing valve may be made as a complete unit which may be dropped into a tubing string and later recovered.

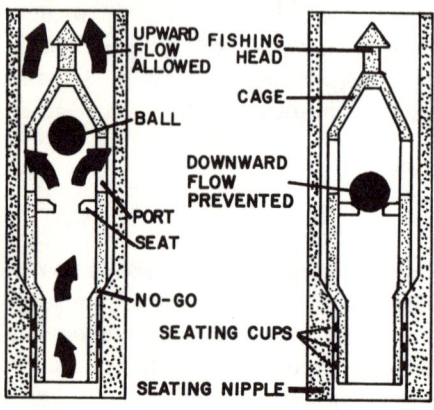

Figure 2-13. A standing valve is used to allow fluid flow into the bottom of tubing but prevent flow out of bottom of tubing.

Rate Control of Flowing Wells

Working bottom hole pressure is the bottom hole pressure when fluid is flowing into the well as opposed to the bottom hole pressure with no flow. If the working bottom hole pressure is more than the sum of the surface pressure and the hydrostatic head of fluid in a well, fluid will move to the surface. Because tubing presents almost no restriction to flow, the rate can be almost unlimited. Fluid is seldom allowed to flow in a wellbore at unlimited rates because there are a number of reasons why flow rate should be limited. Flow is limited to conserve reservoir energy, to improve the efficiency of production, and to prevent development of conditions which can reduce ultimate petroleum recovery.

It is a physical law that as fluid moves through a medium, there is some friction between the fluid and the medium. As the velocity of fluid increases, so does the friction force opposing flow. Some of the reservoir energy which pushes petroleum to the surface is expended to overcome friction effects. One reason to limit flow rate is to reduce the force of friction which is exerted in the reservoir pores, perforations, and tubing. If reservoir energy is not used to overcome excessive friction, it is available to push petroleum to wells. Although it might take longer to recover all petroleum at intentionally reduced rates, these reduced rates help assure that reservoir energy is still available when the last petroleum is being recovered. In other words, one of the most important reasons for controlling the rate of fluid flow is to conserve reservoir energy.

Fluid may flow through a medium in laminar or turbulent flow patterns. When the fluid velocity is low, the flow pattern is smooth and orderly, much like the flow of a quiet, slow-moving river. This type of flow is called *laminar flow*. However, above certain velocities, a phenomenon called *turbulent flow* takes over in which flow is characterized by swirls and eddys. The transition from laminar to turbulent flow is immediate with a slight change in velocity. The transition point depends on the physical dimensions of the flow path, the viscosity and density of the fluid, and the roughness of the medium. The force of friction is an order of magnitude higher for turbulent flow than for laminar flow. Limiting the flow rate of fluid from a well maintains laminar flow.

Fluid flows through a porous matrix as a result of a difference in pressure from one point to another in the reservoir. Flow rate and differential pressure are directly related. Reducing the rate of flow causes a reduction in the difference in pressure and vice versa.

In some reservoirs the difference in pressure across the matrix in the first few feet from the wellbore is so high that there is drastic cooling in this area. This cooling can cause precipitation of solid scales and paraffins in the matrix which can permanently damage the permeability near the well. Limiting flow rate reduces the difference in pressure and the cooling near the wellbore.

The amount of gas that can be dissolved in oil depends on pressure. When there is a significant reduction in pressure near the wellbore due to high flow rates, dissolved gas can come out of solution. This free gas can fill the pore space near the wellbore, severely altering the saturations and reducing the relative permeability to oil. Gas breaking out of solution can alter the flow pattern from hydraulic to aerated or blowing. Controlling the fluid flow rate limits differential pressure near the wellbore.

In reservoirs acted upon by a bottom water drive a phenomenon called *coning* can occur. Under certain conditions, if the flow rate into wells is very high, the aquifer pushes up to the wells in cones (Figure 2–14) instead of pushing upward evenly throughout the reservoir. The water saturates the matrix in the cones, increases the relative permeability to water, and decreases the relative permeability to oil and gas. Once cones develop, a well may begin to produce water rather than oil and gas. If the well becomes unproductive, very expensive techniques are required to regain productivity, and these techniques are not always successful.

When fluid flows into a well very rapidly, it is very likely that a cyclic flow pattern, such as slugging or heading, may develop. Every time the flow of liquid stops, reservoir energy must be expended to accelerate it back to flowing velocity. Constant acceleration and deceleration of liquid wastes energy that could be used to produce petroleum. By controlling or limiting flow rate, cyclic flow behavior can be stopped or reduced in intensity.

Until a few years ago, petroleum removal rates were limited by state or federal regulations. Each well was given an allowable daily production rate that could not be exceeded. These regulations were originally designed for conservation purposes. Now most allowables have been raised to 100% of a well's capability. However, there are still a few wells in which the production rate must be limited to an allowable rate.

Methods of Control

Control of flowing petroleum wells consists of controlling the rate of fluid flow. Flow can be controlled directly with surface or subsurface chokes.

Chokes are control valves which force fluids to flow with constant rate. Rate control is usually achieved by forcing flow through an orifice.

Positive Chokes. This type of choke is a surface control with a fixed orifice size. Figure 2–15 shows the basic operating parts of a positive choke. Figure 2–16 shows a photograph of an actual choke.

Natural Flow   23

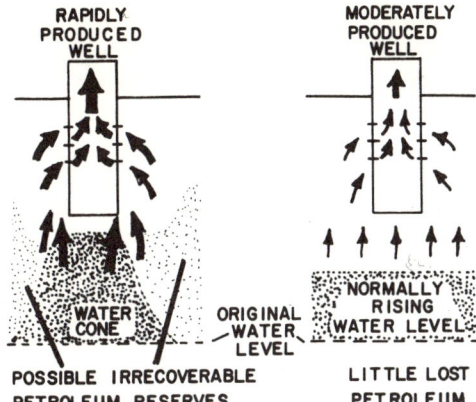

Figure 2-14. Water cones may develop around a wellbore if fluid is allowed to flow into the well too rapidly.

The fixed orifice may be an insert in a master bean (Figure 2–15), or a positive choke bean (Figure 2–16). The orifices in positive chokes are precisely drilled and are available in sizes ranging from $1/64$ inch to about one-half of an inch in increments of $1/64$ inch. The rate at which flow is regulated is determined by the orifice diameter.

As fluids flow through the orifice, the fluids and any suspended solids can erode the insert much like flood waters erode the earth. By using special alloys, beans and inserts can be constructed to a hardness approaching that of a diamond. One of the greatest advantages of positive chokes is that inserts and beans can be made of extremely hard alloys. Another advantage of positive chokes is their low cost. These chokes are constructed of standard-sized precision parts that can be assembled easily, keeping the manufacturing cost low.

The choke setting of a positive choke is fixed and cannot be changed inadvertently. The feature causing this advantage (set orifice sizes in mechanically fixed parts) can also be disadvantageous. As the orifices erode, their diameters increase, and the flow increases above the limiting rate. To compensate for wear or to make a minor flow adjustment, the flow must be stopped, the choke disassembled, a new orifice installed, and the choke returned to service. Although this process is not technically complex, it is time-consuming and requires manpower that could be utilized elsewhere.

Adjustable Chokes. These chokes use handle-operated, cone-shaped stems and seats to form an orifice (Figure 2–17). The orifice size can be increased or decreased by rotating the handle without removing the choke from service. The stem and seat are usually steel alloys that are precisely machined. Locking mechanisms are usually provided which prevent activating the insert to

24 Introduction to Petroleum Production

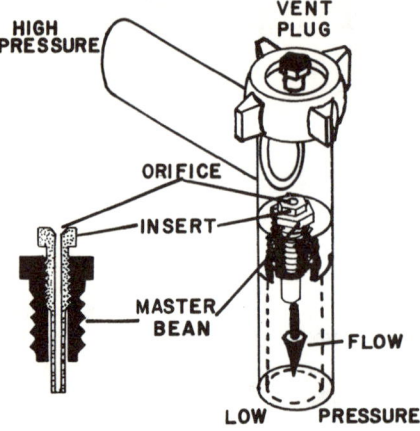

Figure 2-15. Fixed chokes use inserts with predetermined orifice sizes to restrict flow rate.

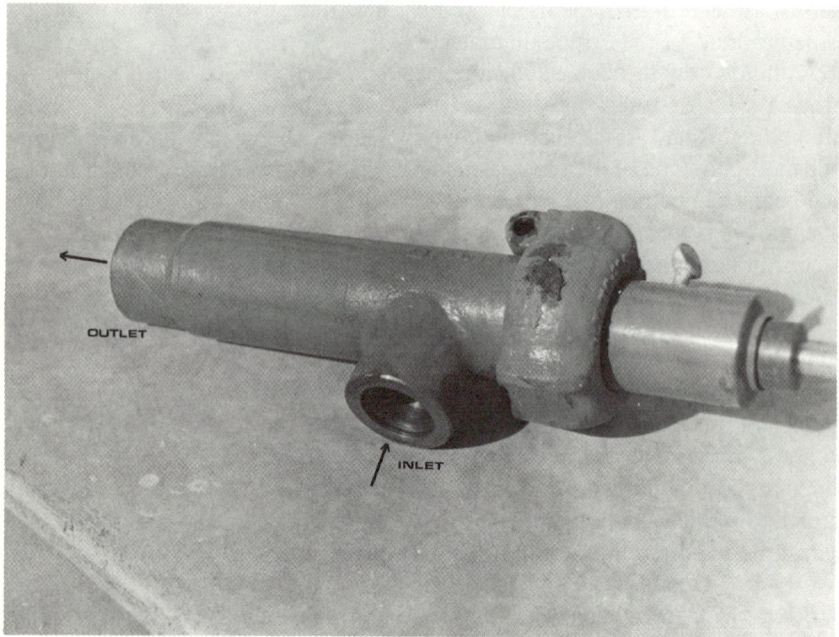

Figure 2-16. A choke restricts flow and reduces pressure at a wellhead or at surface production equipment. (Courtesy of Midessa Equipment Company.)

Natural Flow 25

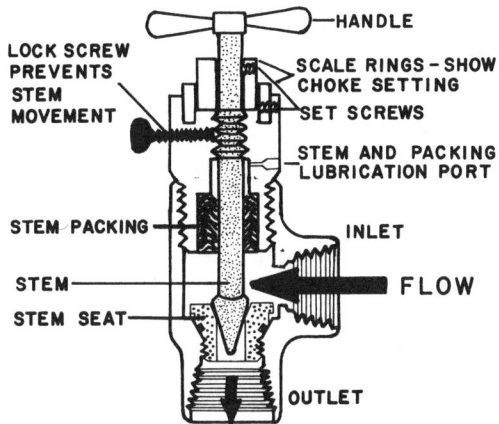

Figure 2-17. Adjustable chokes employ needle valves to allow changing the orifice size.

avoid resetting the choke by accident or by mechanical vibration. The locking mechanism can be deactivated, deleting the protective feature.

One major disadvantage of adjustable chokes is that the orifice size can be changed easily. This, of course, is also one of the primary advantages if the change was intentional.

Bottom Hole Chokes. When tubing cannot be directly exposed to reservoir pressure, bottom hole chokes are installed to reduce pressure before fluid enters the tubing as shown in Figure 2–18. Bottom hole chokes are positive chokes that are designed for subsurface use. They may be equipped with insert orifices, but changing orifices is complicated by the necessity of pulling the tubing to the surface first.

Data Collection for Control of Flowing Wells

The control of flowing wells is a precise operation that depends on knowledge of well and reservoir behavior. The decision to open a choke orifice $1/64$ inch could have long-term effects on the reservoir. Thus, it is necessary to collect, analyze, and retain several data.

The actual flow data, either as instantaneous rates or as volumetric averages, must be collected by metering means. Many producers measure total volume for a lease, while others measure instantaneous rates for individual wells. A change in flow can indicate a change in reservoir conditions, a readjustment of the choke if the change is sudden, or an eroding or plugging orifice if the change is gradual.

26   Introduction to Petroleum Production

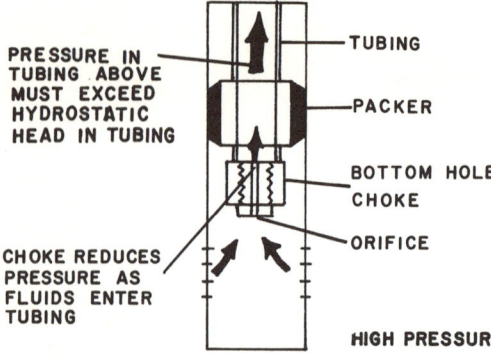

Figure 2-18. A bottom hole choke is used to reduce pressure as fluid enters to protect the tubing.

Tubing pressure of a flowing well is measured with pressure gauges or recorders. A change in tubing pressure usually indicates a variation in reservoir conditions.

The difference between flowing tubing pressure and shut-in tubing pressure (pressures when the well is flowing at normal rate or not flowing at all, respectively) can be used to determine reservoir and subsurface equipment conditions. Shut-in tubing pressure from which the hydrostatic head of fluid is subtracted is the bottom hole pressure. This pressure determines whether or not fluid will flow.

Casing pressure is used to monitor the operation of packers as well as to monitor conditions in the tubing-casing annulus. If there is a sudden increase in casing pressure, the packer seal may have been broken, and work may be required to reseat the packer.

Surface Equipment for Flowing Wells

Production from a flowing well is routed through a pipeline (called a flowline) to a central collection and sales point. Some simple fluid processing is required before the fluid leaves the well site (see Volume 1, p. 144).

Wellhead Heaters

The fluid produced from most flowing wells includes natural gas and some water vapor. As the fluid passes through the choke, the pressure is reduced, and the fluid temperature decreases correspondingly. When fluid moves through a surface flowline, the fluid is cooled by the earth's surface temperature. When a mixture of natural gas and water vapor is cooled, a solid material called *hydrate* can form. Hydrate appears as small crystals much like snow flakes or as a slushy mixture with liquid and is composed of about 90% water and 10% petroleum. The temperature for hydrate formation depends on pressure and may be as low as about 30° F at 200 psi to as high as 70° F at 2000 psi. (Hydrate is not ice, since it may form at temperatures well above the melting point of ice.) Because hydrate is solid, it can plug chokes, wellheads, and flowlines.

Well fluids are heated with wellhead heaters to prevent the formation of hydrates. The heaters use gas-fueled burners to heat fluids.

Direct Heaters. Figure 2–19 shows a direct heater. The fluids are allowed to have direct contact with the outside of the firetube. Direct heaters are the most efficient means of transferring heat to produced fluids.

Petroleum fluids usually contain corrosive chemicals as well as water. The rate of corrosion on the outside of the firetube is accelerated by its high temperature of 700–2000° F. As long as no oxygen is mixed with the petroleum, there is no danger of fire; but if corrosion causes a leak in the firetube, petroleum can enter the firetube, and uncontrolled fire is certain to occur.

Direct heaters require that the firetube and the shell be exposed to the pressure from the well. Because the shell has a large surface area of thousands of square inches, it must be constructed of very thick and expensive steel. Even though the firetube has much less surface area, it must be constructed of equally strong, if not stronger, material to withstand not only the high pressure, but the accelerated corrosion caused by heat. Direct heaters are seldom used for pressures of more than 125 psi, and even with this low pressure, direct heaters are expensive.

Indirect Heaters. Another method of heating fluids is the indirect heater shown in Figure 2–20. High-pressure petroleum fluids are contained within a coil assembly. Heat is transferred to this coil from the firetube via a bath of heat-transfer fluid, such as pure water, salt water (to prevent freezing), or glycol-water mixtures, which allow the bath to be hotter than the boiling point of water. Salt solutions do not usually freeze in cold weather, but scales can precipitate from such solutions and interfere with heat transfer. Pure water does not have scaling tendencies but is subject to freezing when the burner is off. Adding glycol as an antifreeze compound reduces both scaling and freezing problems but decreases heat-transfer efficiency. In every application a heat-transfer liquid must be selected to give optimum operation.

Heat transfer is not as efficient in indirect heaters as in direct heaters, so more fuel must be burned to perform the same heating. The facts that the firetube and shell are constructed for very low pressure, that petroleum cannot leak into the firetube, and that the firetube is not exposed to corrosive fluid make the indirect heater very attractive.

28 Introduction to Petroleum Production

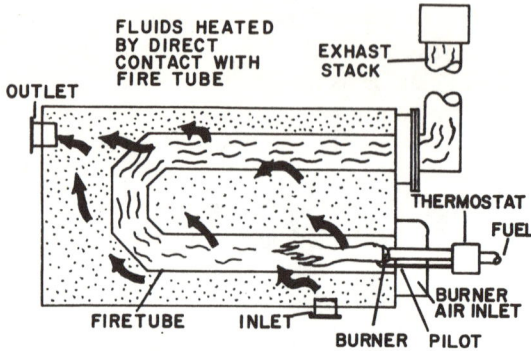

Figure 2-19. High-pressure gas contacts both the firetube and the heater shell in direct heaters.

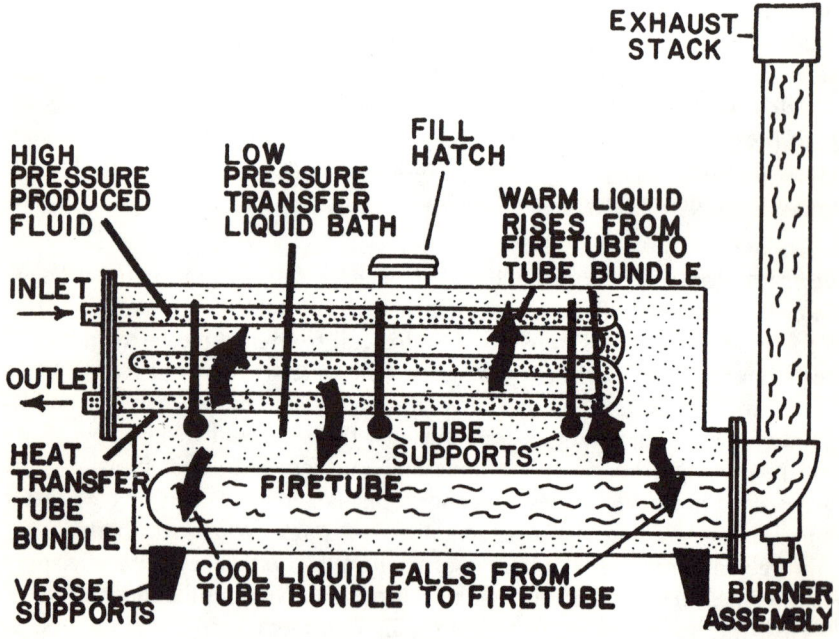

Figure 2-20. High-pressure gas is contained within tubes and does not contact the shell or firetube of indirect heaters.

In the last few years a number of design changes have been made in indirect heaters to improve their heat-transfer efficiency. Figure 2–21 shows an indirect heater which uses multiple-pass coils and a fuel gas preheater coil. Both of these improvements have increased heating efficiency. Other additions to indirect heaters, including extensive use of foam and fiber insulation, have further increased the efficiency of indirect heaters to the point that they are almost as efficient as direct heaters.

Heaters are rated by the amount of heat that is available for transfer from the inside surface of the firetube. Conventional wellhead heater sizes are 100,000–500,000 Btu/hr (British thermal unit per hour). The available heat is determined by firetube size and configuration and design of the burner, which releases heat from fuel gas during oxidation. The best firetube and burner designs furnish about 50% of the available gas energy. The remainder escapes the heater by radiation and convection as the hot exhaust gases leave the stack.

Only 40–50% of the heater's transferable energy actually reaches the produced fluids. The remainder is lost by radiation and convection from the shell. For example, only about 40,000–50,000 Btu/hr reaches the produced fluid from a 100,000-Btu/hr heater. The amount of heat transferred depends on the transfer medium and the rate at which fluid flows through the heater.

The choke used with a heater may be installed in front of the heater. If the choke is installed below the heater, as shown in Figure 2–22, any choke may be used. The heat is applied to the fluid so that after it cools through the choke, it is still warm enough to prevent the formation of hydrates. When the choke is below the heater, the full tubing pressure is applied to the heater, requiring high-pressure coils in indirect heaters.

If the choke is installed in front of the heater, as shown in Figure 2–23, the choke must be a long-nosed choke (if it is an adjustable choke) so that the pressure drop in the choke occurs inside the heater. If this arrangement were not used, hydrates could form in the choke orifice before the fluids could be heated. The advantage to this configuration is that the normal operating pressure of fluid is significantly lower than before; however, the pressure rating of the coil must be for the maximum tubing pressure, because, if flow is blocked downstream of the heater, full pressure occurs regardless of the choke setting. Another advantage to installing the choke in front of the heater is that fluids do not have to be as hot to prevent hydrate formation, and significant fuel reductions can be realized.

Heaters are often mounted on skids (along with the appropriate chokes, fuel gas connections, and heater controls) as a complete assembly called a production unit (Figure 2–24). In harsh environments it is necessary to supply the production unit in an insulated and sometimes heated enclosure as pictured in Figure 2–25.

30   Introduction to Petroleum Production

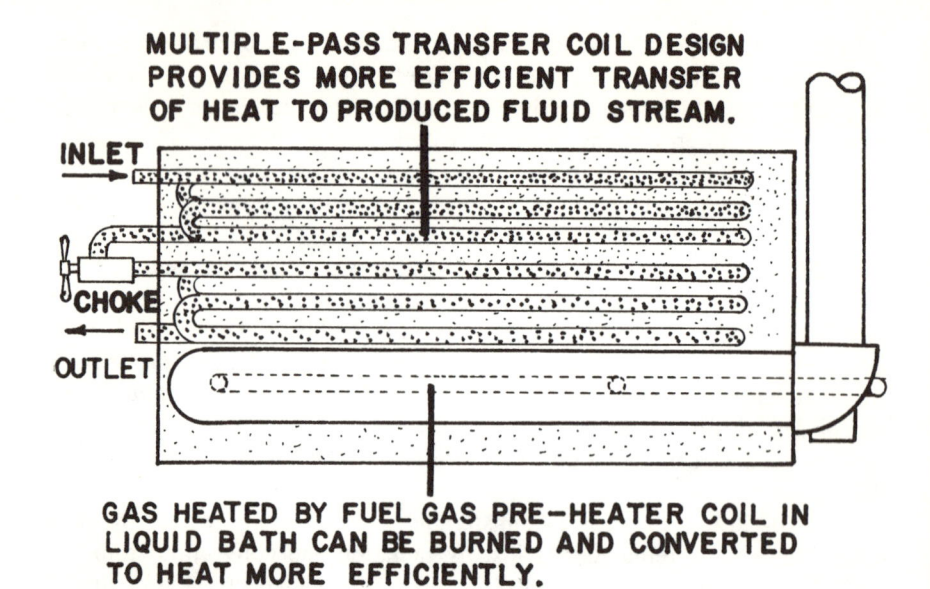

MULTIPLE-PASS TRANSFER COIL DESIGN PROVIDES MORE EFFICIENT TRANSFER OF HEAT TO PRODUCED FLUID STREAM.

GAS HEATED BY FUEL GAS PRE-HEATER COIL IN LIQUID BATH CAN BE BURNED AND CONVERTED TO HEAT MORE EFFICIENTLY.

Figure 2-21. Recent design improvements in indirect heaters have improved heating efficiency.

Two-Phase Flow

In most flowing wells fluids in both the gaseous state (or phase) and liquid phase travel simultaneously. This is called two-phase flow. Fluid flowing in two phases has the properties of viscosity and density which affect the presence of turbulent or laminar flow and friction pressure losses, but the properties of a two-phase stream are different from those of the gas and liquids composing the stream. At best it is often difficult to predict or control the behavior of a two-phase stream precisely.

Measurement of the flow rates of gases and liquids from a well is vital for accounting for petroleum production as well as for controlling the flow rate. Unfortunately, meters capable of measuring both gas and liquid simultaneously are not available for petroleum production applications because they are far too delicate and expensive. Inexpensive and durable meters that can measure gas or liquid, but not both, are available for use.

To contol and measure the fluid streams from a flowing well adequately, it is necessary to divide them into their component gas and liquid streams. Vessels, called two-phase separators, are used for splitting the phases. Separators are described in detail in Chapter 5; however, for the present discussion, it is enough to know that these are mechanical devices which take

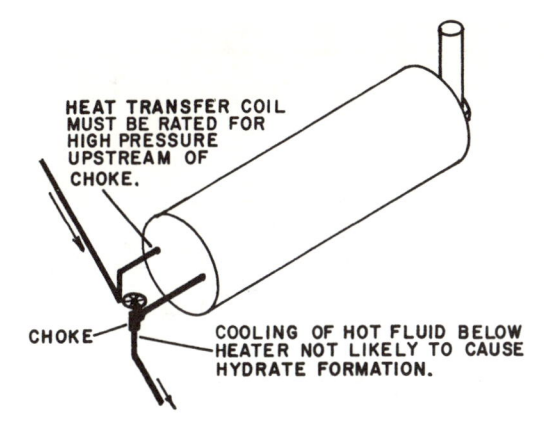

Figure 2-22. Reducing pressure with a choke after gas has been heated is not likely to form hydrates.

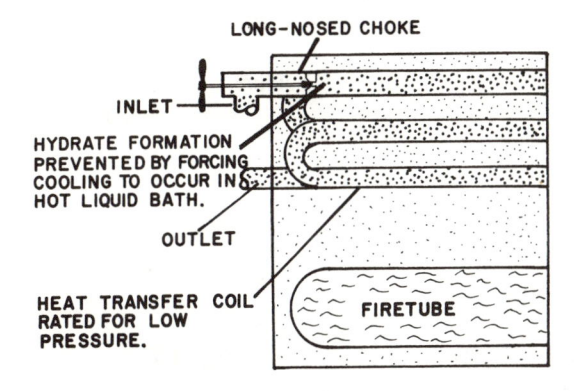

Figure 2-23. Reducing pressure ahead of a heater normally requires a long-nosed choke.

advantage of the force of gravity to allow gas to leave the separator by one path as liquids leave by another.

The gas used for fuel in heater burners, as well as the gas used for operating pneumatic instruments, must be dry. Petroleum liquids mixed with fuel gas cause unpredictable burner operation and control.

Figure 2-26 shows a scrubber which is used to remove most oil and water from a fuel gas stream. A scrubber simply allows liquid to fall from a gas stream by forcing the incoming fluid to follow a serpentine path. A scrubber may be equipped to drain accumulated liquid automatically or to stop flow until liquid can be drained manually as shown in Figure 2-27. A scrubber is not normally considered a separator, since it accumulates liquid instead of furnishing separate flow paths for liquid and gas.

32 Introduction to Petroleum Production

Figure 2-24. A production unit contains a heater, fuel equipment, and a choke to control the production from a gas well. (Courtesy of Engelman-General, Inc.)

Figure 2-25. An enclosed production unit has a building to retain internal heat and can be used in cold climates. (Courtesy of Midessa Equipment Company.)

Natural Flow 33

Figure 2-26. A fuel-gas scrubber eliminates liquid from the gas used for burner fuel as well as that used for pneumatic instruments and controllers. (Courtesy of Engelman-General, Inc.)

Fluid Flow Measurement

In the petrochemical industries there are analyzers that are capable of monitoring a fluid stream, analyzing its components, and giving an indication of the flow rate of the constituent parts. These are complicated electronic and mechanical instruments which often use small computers. Unfortunately, these instruments are very delicate, require a high technical level for maintenance and operation, and are extremely expensive. Such instruments are not usually useful for measuring the fluid flow rates from petroleum wells.

There are a number of simple, durable, and inexpensive meters which are applicable for fluid measurement in production activities. These meters are capable of measuring only single-phase flow of liquid or gas.

Orifice Gas Meters. Use of orifice gas meters is one of the oldest and most widely accepted methods of measuring gas flow rate. Figure 2–28 is a diagrammatic view of the orifice used for such metering. The principle involved is: if gas is forced to flow through an orifice of known size and with a constant pressure downstream of the orifice, the flow rate is directly proportional

34 Introduction to Petroleum Production

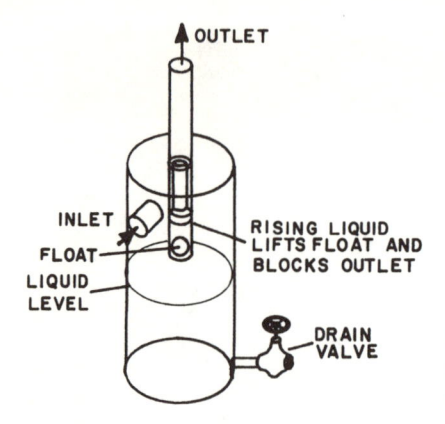

Figure 2-27. Scrubbers can be equipped to prevent flow when the scrubber fills with liquid.

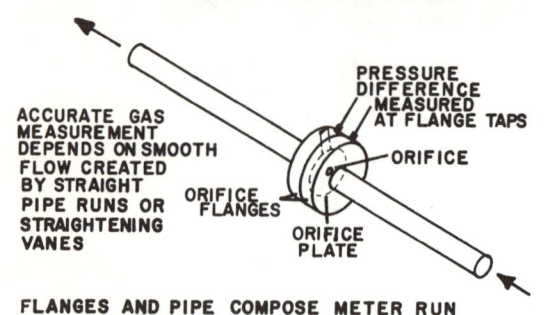

Figure 2-28. Gas flow rate can be determined from the pressure and difference in pressure across an orifice.

to the square root of the product of the pressure and the difference in pressure across the orifice (the differential pressure). The constant of proportionality depends on non-variable characteristics of the metering system such as the orifice size, the inside diameter of the pipe, the position of the taps; and on variable characteristics such as gas viscosity, density, compressibility, and, to a small extent, temperature. Assuming that the composition of gas remains constant and thus the viscosity, density, and compressibility of the gas remain constant, the flow rate of gas is measured through an orifice simply by calibrating the metering system and measuring pressure and differential pressure.

The relationship of many parameters makes the measurement of gas through an orifice system seem quite complex. In practice, orifice measurement can be simplified to give reasonably accurate measurements (within 1% of true value). More accurate measurements are possible but are seldom re-

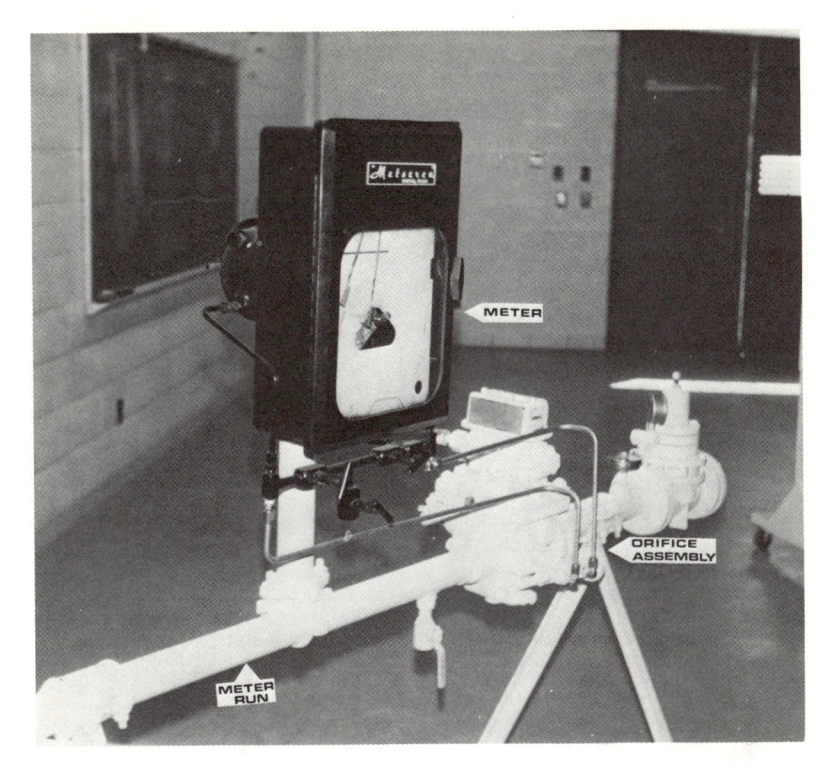

Figure 2-29. A meter run consists of long, straight sections of pipe before and after the orifice plate so gas turbulence is minimized. (Courtesy of Midland College.)

quired. Figure 2–29 is a photograph of the meter run (the straight pipe leading to and from the orifice) containing the orifice flanges for a paddle orifice plate shown in Figure 2–30. Figure 2–31 is a cutaway of the orifice assembly for another form of the orifice plate.

Mercury Manometer Orifice Meter. Figure 2–32 shows the mechanical linkages used to operate a mercury manometer gas meter. The difference in pressure causes vertical movement of a column of mercury upon which a steel float rests. This float drives the differential pressure pen across a circular chart rotated by a mechanical clock. The downstream pressure (called static pressure) is connected to a Bourdon tube which, in turn, drives the static pen on the chart. These two pens record the variation of pressure and differential pressure on a chart pictured in Figure 2–33.

The pressure and differential pressure are recorded on a square root scale so that by reading the two pressure values and by multiplying the product of

36 Introduction to Petroleum Production

Figure 2-30. A paddle orifice plate is simply a flat plate with an extension to facilitate manipulation. (Courtesy of Midland College.)

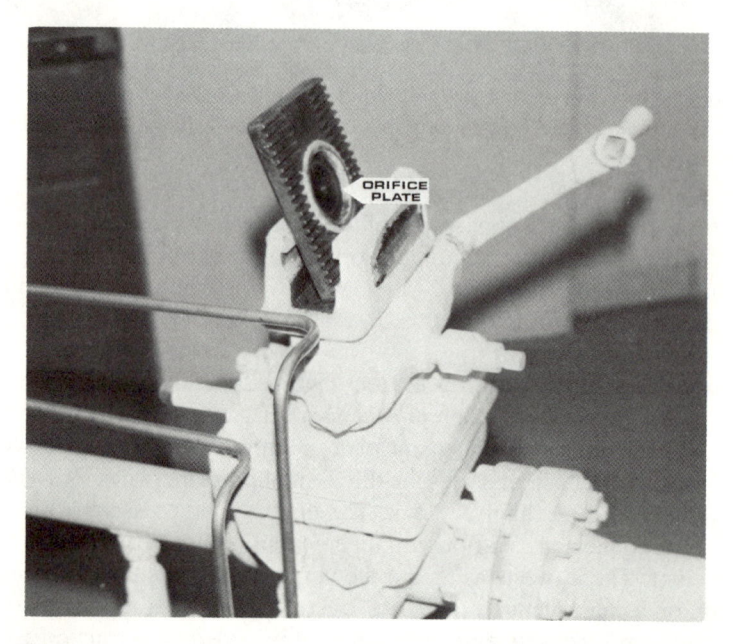

Figure 2-31. This orifice assembly allows cranking the orifice plate in and out without danger of leakage. (Courtesy of Midland College.)

Natural Flow 37

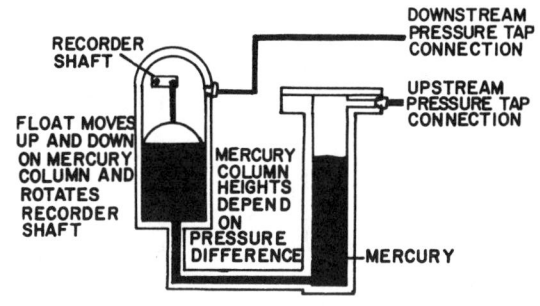

Figure 2-32. A mercury manometer gas meter uses slight changes in depth of a column of liquid mercury to operate the pens of a recorder.

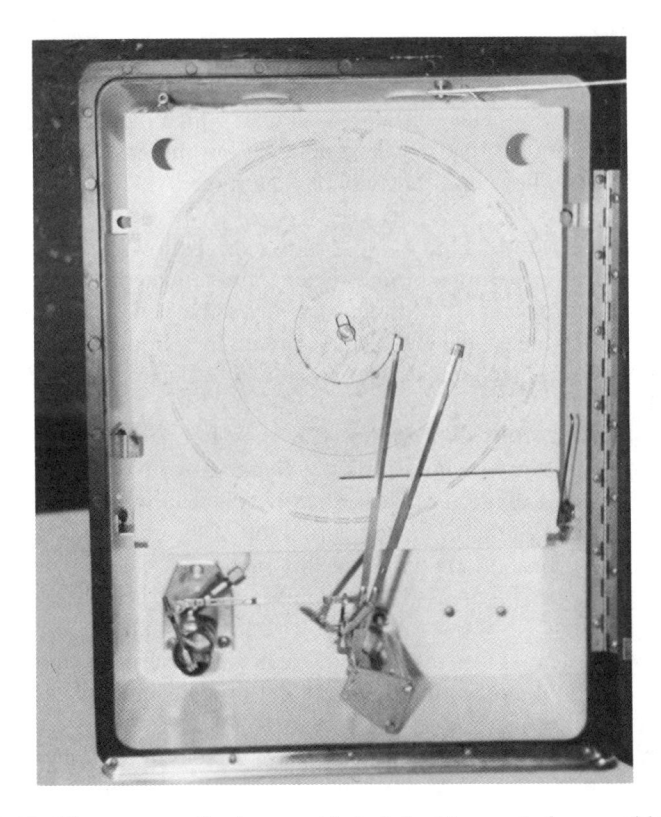

Figure 2-33. The mercury float assembly is linked to a set of pens which ink the value of pressure and differential pressure on a circular chart. (Courtesy of Midland College.)

38 Introduction to Petroleum Production

DIFFERENTIAL PRESSURE CONNECTIONS

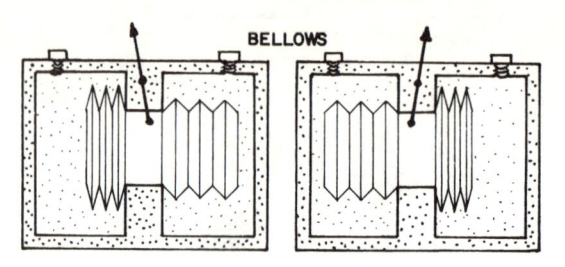

BELLOWS

CHAMBERS ARE NOT CONNECTED, AND GAS DOES NOT FLOW BETWEEN THEM.

Figure 2–34. A bellows gas meter uses expansion and contraction of a bellows to operate recorder pens and measure gas flow.

these two values by a meter constant, a number which includes the effects of all other parameters for that metering application, the gas flow rate in standard cubic feet (scf) per unit time can be determined.

Bellows Orifice Meter. Figure 2–34 shows the bellows used to drive the differential pen in a bellows orifice meter. Differential pressure causes the brass bellows to expand or contract and move the pen. Like the mercury manometer orifice meter, the static pen is driven by a Bourdon tube while the chart is moved by a mechanical clock as shown in Figure 2–35.

Positive Displacement Gas Meter. Gas may be measured by making it move through a meter which has moving chambers of fixed volume (Figure 2–36). The speed of rotation of these chambers is determined by the volume of gas passing through them. Since the rotation of the chambers is caused by fixed volumes of gas, the meter using this method is called a positive displacement gas meter (Figure 2–37 is a cutaway view). The chambers through which the gas moves are the cavities between the impellers. As gas flows, the impellers rotate and turn a gear train connected to a meter (much like the odometer on an automobile) that registers gas flow in standard cubic feet.

Inferential Gas Meters. Gas flow can also be measured by devices called inferential gas meters. Such meters do not measure the actual volume of gas as positive displacement meters do, nor do they measure the physical characteristics of the gas flow stream as does the orifice meter. Rather, inferential

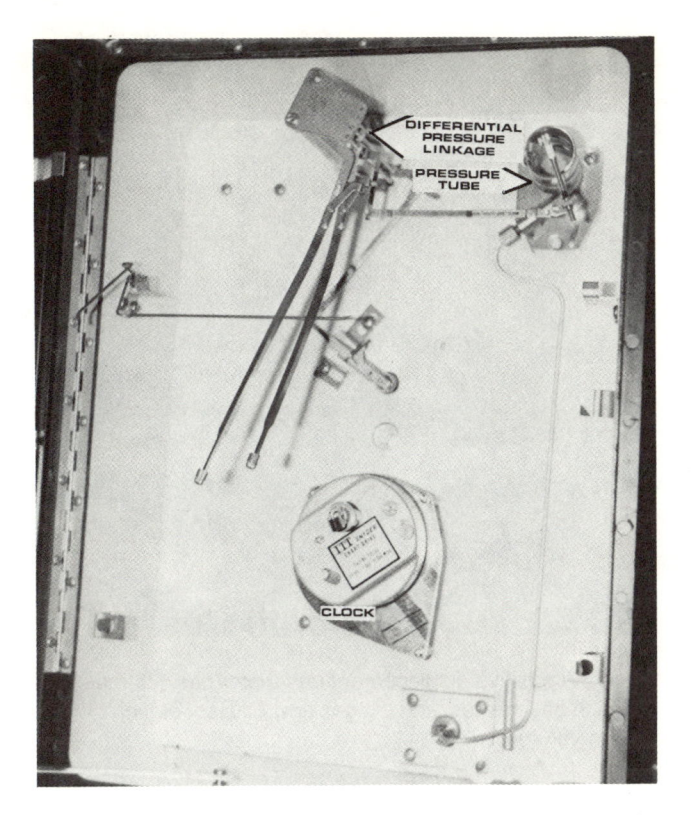

Figure 2-35. The interior of a bellows orifice meter is a simple arrangement of pen linkages with space for the mechanical clock which rotates the chart. (Courtesy of Midland College.)

meters measure the effect of the impact of the mass of moving gas on a mechanical system.

One of the commonly used inferential gas meters is the turbine meter pictured in Figure 2–38. Gas rotates the turbine blades. A magnetic coil senses the passage of the blades under it and sends electric impulses to an electronic counter. This counter is calibrated to take into account the meter size, the blade shape, temperature, density, and other parameters so that the counter records gas flow in scf.

Another type of gas turbine meter uses the turbine to drive a gear train connected to an odometer. Again, gas flow is recorded in scf.

Positive Displacement Liquid Meters. Liquid flow may be measured with positive displacement meters which employ the same principle used for gas

40 Introduction to Petroleum Production

Figure 2-36. A positive displacement gas meter has volume chambers through which gas passes, and the gas flow causes rotation of a counter to show gas volume.

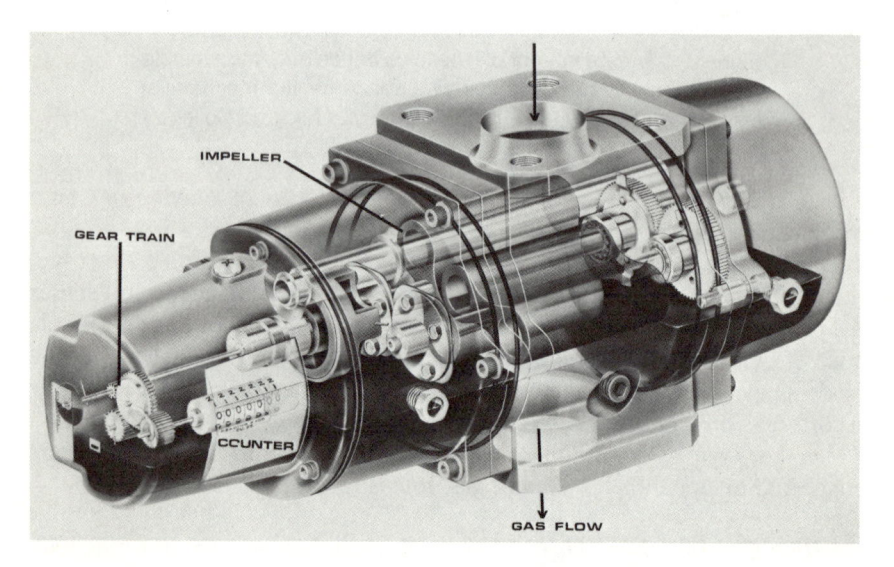

Figure 2-37. The impellers and gear train of a positive displacement gas meter are precisely machined to allow smooth, friction-free movement as light gases flow through it. (Courtesy of Dresser Measurement Division.)

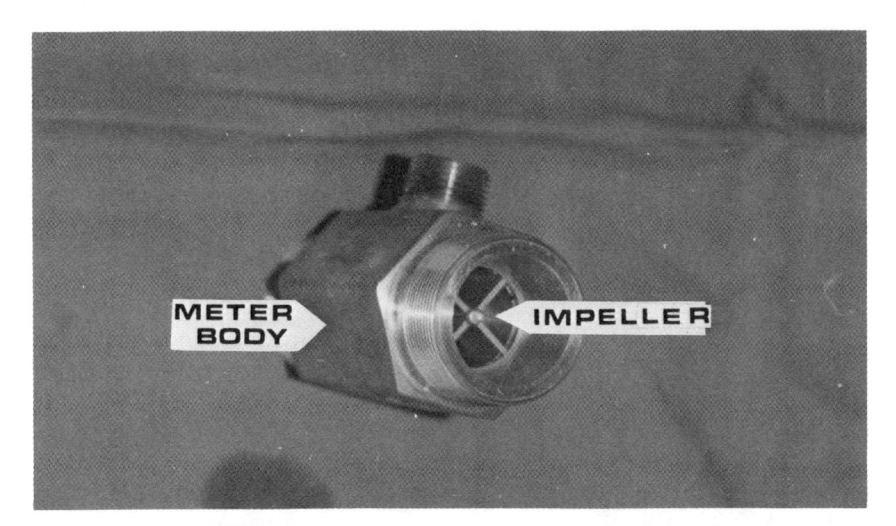

Figure 2-38. Gas moving through a gas turbine meter turns turbine blades whose motion is detected electronically. (Courtesy of ITT Barton.)

measurement. Because liquid is more dense and incompressible, the spacing between meter parts must be greater than for gas meters. The gear trains must also be different because liquid flow is measured in gallons or barrels (1 barrel = 42 gallons). In positive displacement gas meters lubrication is a major consideration, while lubrication is of little importance with positive displacement liquid meters. The liquids being measured act as the lubricant for these meters.

Inferential Liquid Meters. The least expensive liquid meters are inferential meters. Again, these meters measure the effect of mass movement rather than detect the actual volumetric movement of liquid.

One type of inferential liquid meter is the paddle meter shown in Figure 2–39. Liquid pushes the paddles and rotates the meter, which in turn rotates a gear train and an odometer which records liquid flow. This particular meter also uses the paddles to rotate a sample port. At set increments of rotation, the port is opened, and a small volume is released into a sample container. At the conclusion of the metering period, that sample may be analyzed to determine the average composition of the liquid stream.

Another type of inferential liquid meter is a turbine meter shown in Figure 2–40. A liquid turbine meter differs from a gas turbine meter in the number and shape of blades. More important, however, is the fact that liquid turbine meters must use special thrust bearings to support the turbine to avoid damage by liquid impact.

A vortex meter is a newer liquid meter which uses the principle that when liquid flows past an obstacle, vortices or eddies are formed just as a flag

42 Introduction to Petroleum Production

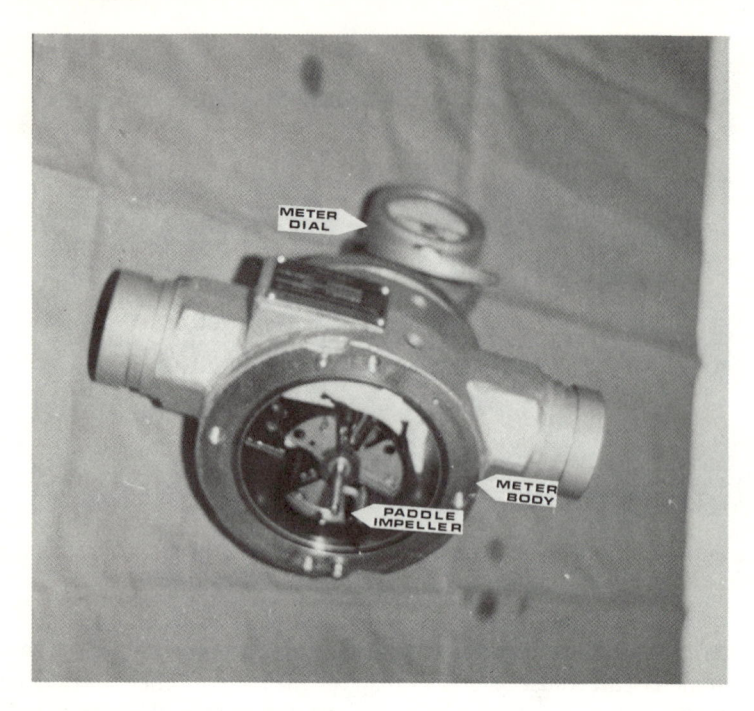

Figure 2-39. Liquid pushes the paddle impeller and turns the meter dial. (Courtesy of ITT Barton.)

ripples as wind passes it. This meter uses special electronic devices and circuits to count these vortices. The fluid flow rate is proportional to the rate at which the eddies are formed.

Liquid Orifice Meters. Liquid can also be measured by gauging the differential pressure across an orifice through which liquid flows. This is an accurate method of measurement, but it requires reading a chart to determine the actual rate.

Two-Phase Measurement

Although the principles for gas and liquid measurement are similar, the meters built to apply them differ significantly for gas and liquid. One reason is that gas meters must be more delicate to measure the lower-density fluid; however, the impact of gas on these meters is not as great as that of liquid, so gas meters need not be as sturdy as liquid meters. Another difference is that gas cannot lubricate moving parts, and an oiling method must be provided. However, liquid (even water) can be used to lubricate liquid meters.

Figure 2-40. Liquid turns the turbine blades of a liquid turbine meter while an electronic detector measures the rate of liquid flow.

Gas meters are calibrated for low density. Liquid in a gas meter might not cause immediate damage, but the heavier fluid would not be measured accurately. Thus, gas and liquid meters are not interchangeable.

Petroleum Recovery by Natural Flow

In theory, petroleum can be recovered from a reservoir by simply drilling a well and allowing fluid flow. In practice, however, natural flow recovery requires careful consideration and planning of the subsurface and surface equipment used for recovery. The flow-rate control method is the most important single consideration because it determines not only the rate of recovery but also how the reservoir energy is used. Too rapid production can prematurely exhaust reservoir energy or permanently damage the reservoir. Too slow production defers the supply of the vitally needed energy resource and slows financial recovery from the very large capital expenditure made to drill and complete the well.

No matter how effectively the reservoir energy is utilized, the reservoir pressure will decrease to the point that fluids will not reach the surface. If this occurs when recoverable petroleum remains, some method must be used to lift the remaining liquid. Energy must be supplied from surface sources to lift the liquid using artificial lift methods, the subject of the next chapter.

Chapter 3
Artificial Lift

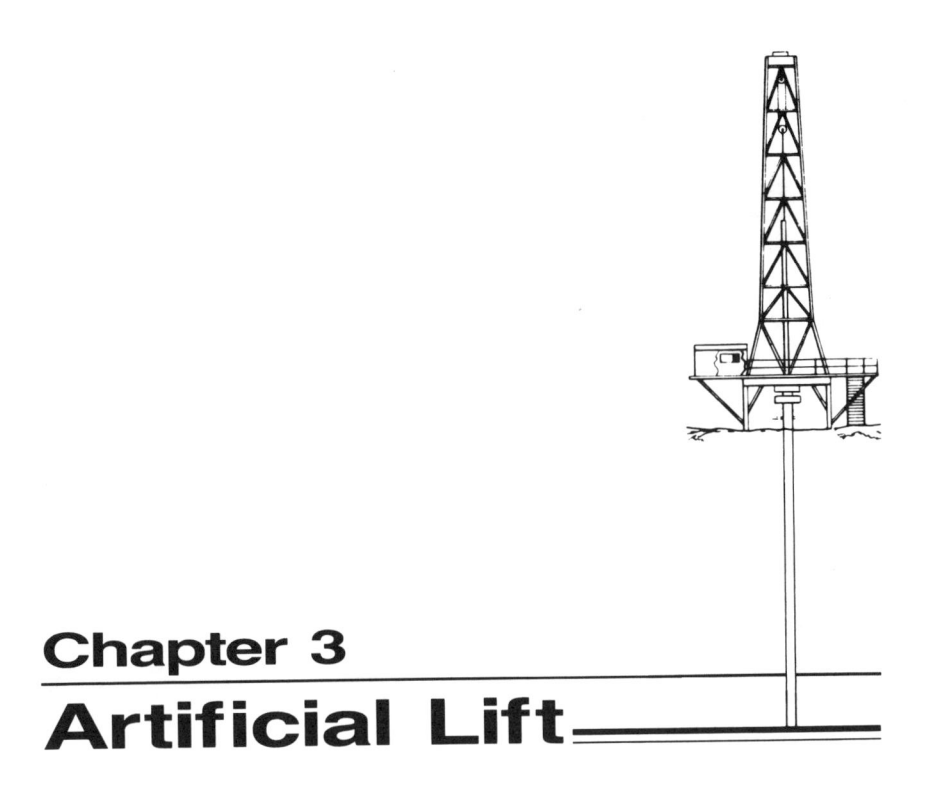

If there is not enough energy in the form of reservoir pressure to overcome surface and hydrostatic pressure, petroleum liquids will not flow to the surface regardless of the amount of petroleum in the reservoir. It is then necessary to use artificial lift to recover the petroleum. Stated simply, artificial lift involves methods of transmitting energy to the bottom of a well to supplement natural energy in pushing liquids to the surface.

Artificial lift must be used when liquids will not flow to the surface naturally; however, it may also be used when the natural flow rate is not as high as desired. Artificial lift can also be used to increase flow rate, since high bottom hole pressure is not required to lift liquid and can be reduced. In some gas production applications water enters the wellbore, develops a hydrostatic head opposing gas flow, and interfers with gas production. This impediment to gas flow may be overcome by using artificial lift to remove water from a wellbore in an operation called *dewatering*.

The decision to use artificial lift must be based on economic factors. The cost of equipment used for artificial lift is high, but this cost may be more than offset by the increased production rates possible. In fact, artificial lift is often used so that the production rate can be increased when a well is still capable of flowing.

46 Introduction to Petroleum Production

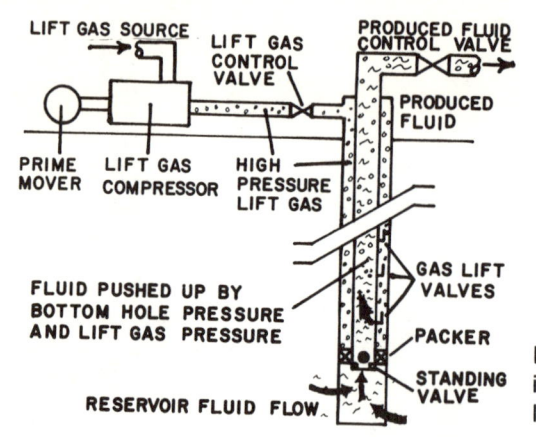

Figure 3-1. Gas lift uses energy in the form of compressed gas to lift produced fluid to the surface.

There are several types of artificial lift methods, including

1. Gas lift
2. Plunger lift
3. Sucker rod pumping
4. Electric submersible pumping
5. Hydraulic pumping

The principles of these methods differ vastly, but they all require transmitting energy in some form to the bottom of a well (See Volume 1, pp. 141-143).

Gas Lift

Gas lift is a means of artificial lift whereby energy is transmitted to the bottom of a well in the form of compressed gas. Gas lift is shown schematically in Figure 3-1 and requires pumping natural gas down the tubing-casing annulus with a compressor, injecting the gas into the tubing through pressure-operated gas lift valves, and using the expansion of the gas (caused by pressure reduction as it passes through the valves) to assist in pushing liquid to the surface.

Gas lift can be used in many oil wells, and it is often used to dewater gas wells. Gas lift is usually used when the reservoir pressure is still high, and when a significant volume of gas is available. It is often used as a means of assisting the natural flow of a well by aerating the fluid column to decrease its hydrostatic head.

Artificial Lift   47

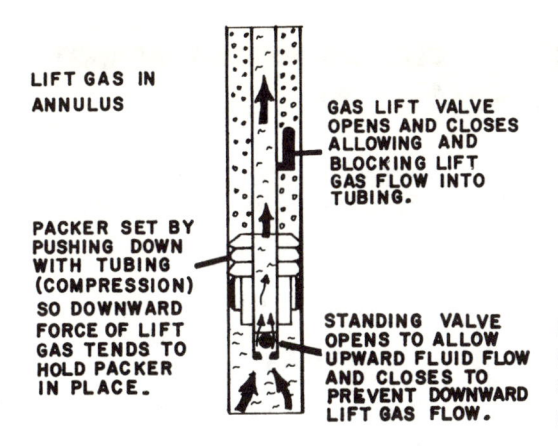

Figure 3-2. A well equipped for gas lift usually has a packer and standing valve as well as gas lift valves.

Subsurface Equipment

Figure 3-2 shows the subsurface configuration used in gas lift systems. Tubing of the same type used for flowing wells is used. Because pressure differentials across the tubing can sometimes be great, it may be necessary to use high-yield tubing; H-40 and J-55 tubing are usually satisfactory.

The pressure used for gas lift is high enough to push gas into the reservoir, and a packer must be used to isolate the annular pressure from the lower bottom hole pressure. The difference in pressures above and below the packer exerts considerable force to unseat the packer by making its packing elements and slips lose their grip on the casing. This force is exerted downward, so the packer may be set in compression to overcome this force; otherwise, the packer may be set with enough tension to hold the packer and overcome the unseating force.

Standing Valve. During the process of gas lift operation, the pressure inside the tubing can be very high. A standing valve is placed in the bottom of the tubing to keep from pushing lift gas and liquids back into the formation. However, when the tubing pressure drops, the standing valve opens and allows fluid to enter the tubing.

Gas Lift Valves. Figure 3-3 is a schematic view of the subsurface valve used for gas lift. The dome is charged with gas prior to attaching the valve to the tubing. The dome pressure is chosen to cause the valve to open at a specific combination of annulus and tubing pressure.

48 Introduction to Petroleum Production

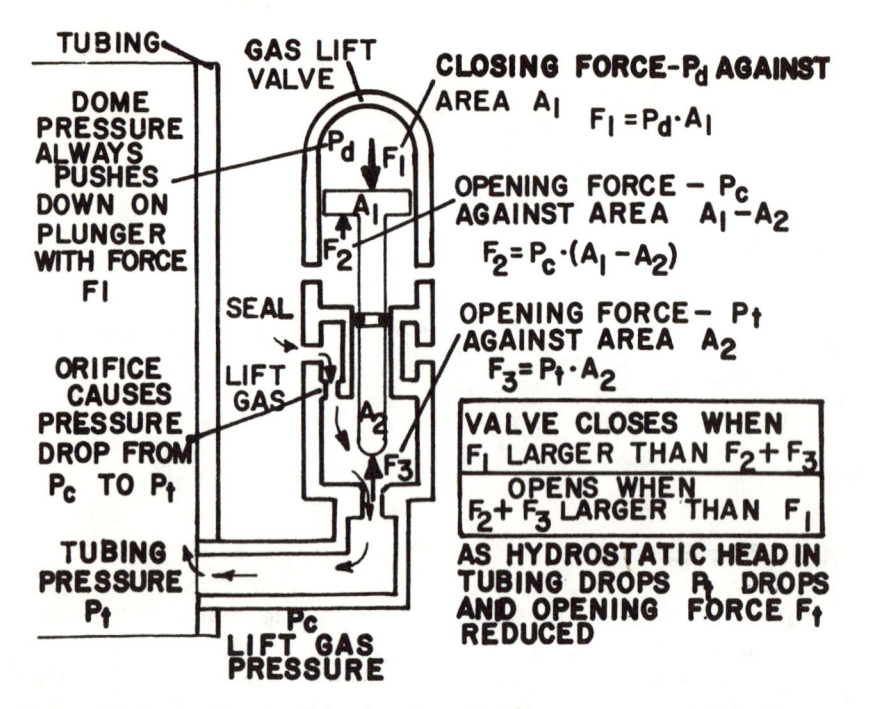

Figure 3-3. A gas lift valve operates when different pressures act against surfaces in the valve.

When the gas lift valve is attached to tubing and placed in a well, the bellows is subjected to forces caused by the dome, casing, and tubing pressures. When the forces acting to open the valve (Figure 3–3) exceed those acting to close it, the valve opens, allowing gas to flow through an orifice in the valve and into the tubing.

A cutaway view of a gas lift valve is shown in Figure 3–4. This is called a pressure-operated gas lift valve. An early valve design was the pressure-differential valve, which used another arrangement of ports and pistons but was similar in operation. The pressure-differential valve has been replaced by the pressure-operated valve in most installations.

Principles of Gas Lift

There are two methods of applying gas lift to producing wells. One is called continuous gas lift. In this method gas is injected into the tubing at all times and is used more for aerating the fluid column than for actual lifting.

Artificial Lift   49

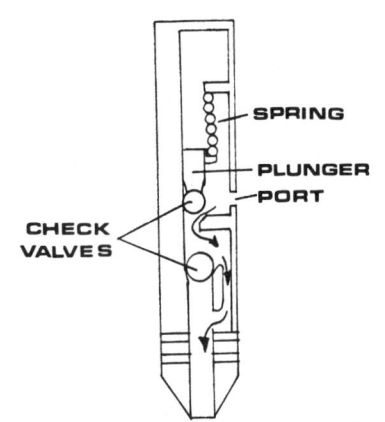

Figure 3-4. A pressure-operated gas valve stays open as long as tubing pressure exceeds the spring tension and dome pressure in the valve.

The other method is called intermittent gas lift, and it requires injecting gas for a time, unloading the tubing, then stopping gas injection while fluid refills the tubing.

Continuous Gas Lift. Gas is supplied at constant pressure in the annulus, and the gas lift valve near the bottom of the tubing allows gas into the tubing. The dome pressure is low so that low casing pressure may be used.

Intermittent Gas Lift. This method is pictured in Figure 3-5. Reservoir pressure pushes fluid into the tubing through the standing valve. At regular intervals determined manually or automatically by a timer, an injection gas valve at the surface opens. The dome pressure of the valve is set to open the valve fully, and a slug of gas is injected below the fluid. As gas injection continues, the liquid is pushed to the surface. Gas injection is stopped when the surface valve is closed, and the standing valve which closed when the gas lift valve opened allows fluid to enter the tubing. The cycle is repeated often enough to lift as much fluid as enters the tubing.

Emptying the Annulus. When the tubing is first placed in a well, the annulus may be partially or completely filled with liquid. When a well's gas lift is stopped for an appreciable time, the tubing fills with liquid, and the gas lift valves open to allow liquid to enter the casing. Although liquid does not harm gas lift valves, it interferes with the operation of the gas lift system. It is necessary to remove annular fluid in a process called *unloading*.

A well is often equipped with several valves that are used only for unloading the annulus as shown in Figure 3-6. To unload the annulus, gas injection

50 Introduction to Petroleum Production

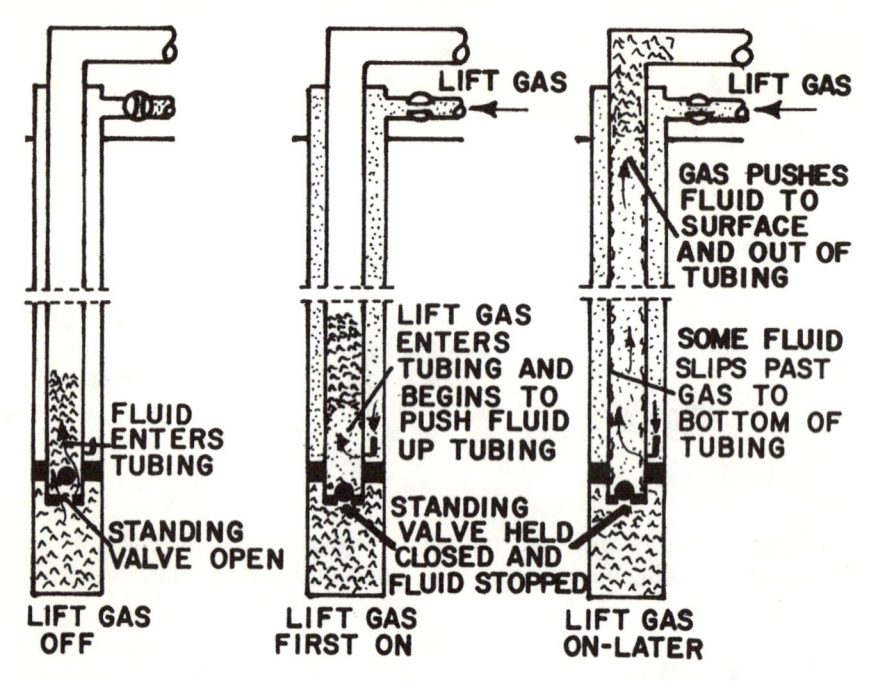

Figure 3-5. The intermittent gas lift method consists of injecting gas down the annulus, lifting fluid for a time, and then stopping gas injection for a time.

is started, and all valves open. Gas begins to enter the tubing through the top valve and pushes the liquid down and into the tubing through the other valves as shown in Figure 3–7. As gas injection continues, gas enters the tubing through lower valves and pushes the liquid down until the annulus has been emptied.

Gas Source

Before gas lift can be used, a source of lift gas must be secured. In most cases a portion of the natural gas produced from the reservoir is used for lift gas. It is separated from liquids and used for injection. When there is not enough gas produced for gas lift, the additional gas may be purchased or transferred from another lease.

Gas Compression

When the volume of gas is reduced, the pressure increases. This physical law is the basis for compressors used to provide high-pressure gas for lift.

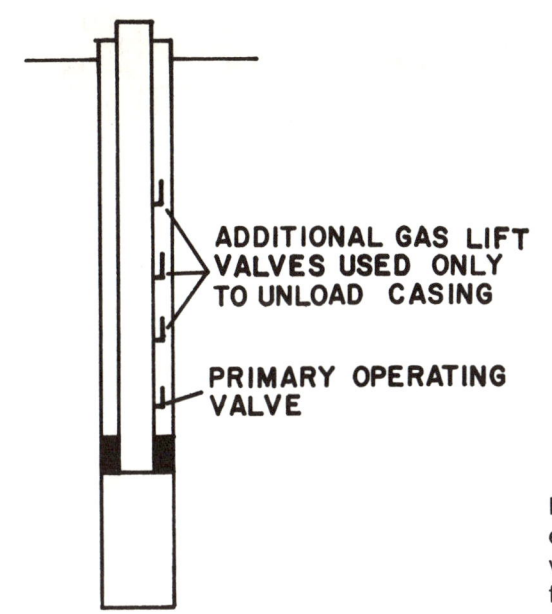

Figure 3-6. A well may be equipped with several gas lift valves to facilitate removing fluid from the annulus.

Figure 3–8 shows a positive displacement gas compressor. This is the most commonly used compressor for gas lift, and this and other types of compressors are discussed in more detail in Chapter 5. A compressor is driven by a prime mover which may be an electric motor or an internal combustion engine using natural gas for fuel.

Wellhead Equipment

Figure 3–9 is a sketch of the wellhead arrangement for a well using gas lift. When gas lift is used, there may be two chokes: one reducing the pressure of the lift gas before it enters the casing, and the other reducing pressure of tubing fluids as they enter the flowline.

Another valve used to control the flow of lift gas is a pressure regulator, shown diagrammatically in Figure 3–10. This valve is intended to maintain a constant pressure downstream of the valve and can be used instead of a choke.

Valves must be used to provide for isolation of the tubing and casing as well as completely blocking the compressed gas flow. Valves are sometimes provided to isolate chokes and pressure regulators from high pressure so they can be serviced.

52 Introduction to Petroleum Production

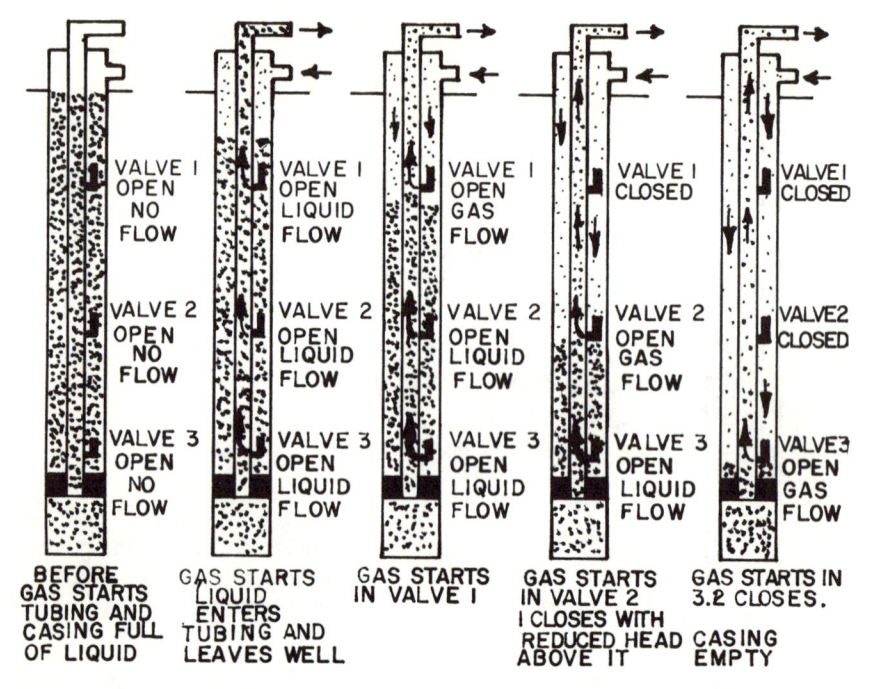

Figure 3-7. Several valves open and close sequentially in the process of emptying fluid from the annulus.

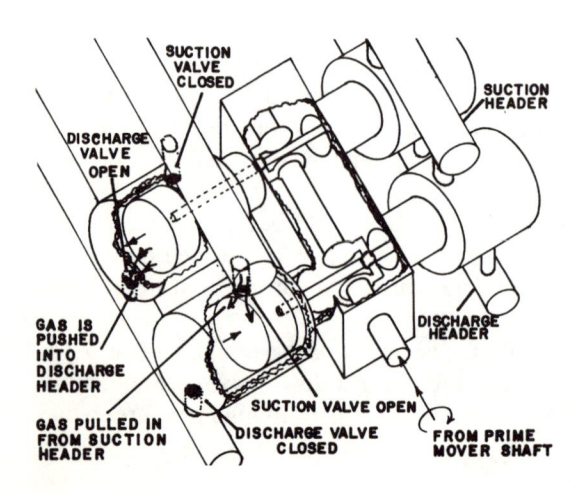

Figure 3-8. A positive displacement compressor is used to pressurize the gas used for gas lift.

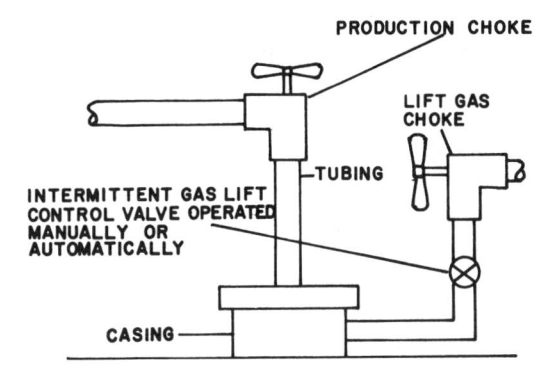

Figure 3-9. The wellhead arrangement of a well produced with gas lift may include two chokes, an intermittent valve, and several general-purpose valves.

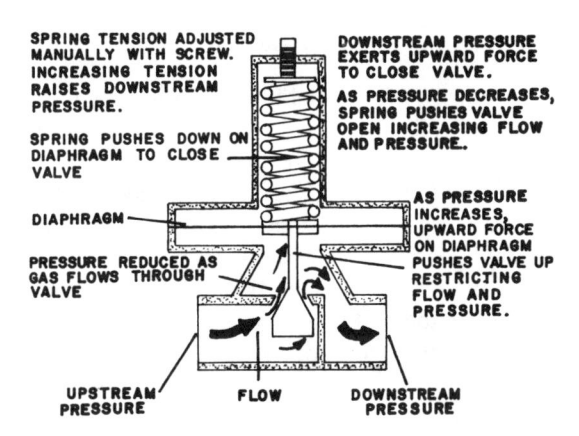

Figure 3-10. A pressure regulator may be used instead of a choke to control lift gas pressure at the wellhead.

Applications of Gas Lift

Gas lift is usually one of the least costly methods of artificial lift when many wells must use the compression facilities; however, it is one of the least efficient in terms of energy usage. As liquid is lifted in tubing, some of the liquid falls back toward the bottom; this loss of fluid is called slippage. Significant slippage requires more lift gas than is needed to lift fluid just once from the bottom of a well and is more pronounced in intermittent gas lift.

Since continuous gas lift aerates liquid columns while allowing bottom hole pressure to lift the fluid, it is usually restricted to shallow wells less than 5000 feet in depth. The gradient of the liquid column can only be limited by injecting gas. At more than 5000 feet, continuous gas lift does not assist reservoir pressure enough to lift liquids. To lift liquids from great depths, the injection pressure would have to be so high that the standing valve would never open, and fluid would never flow into the wellbore.

The selection of equipment size and use in gas lift is affected by the well's productivity, proration, flow rates, and costs. The productivity of a well determines the rate at which fluid enters the tubing and affects the amount and pressure of the lift gas and the valve pressure selection. Proration (government regulation of flow rates) and conservation of reservoir energy may reduce the anticipated flow rate and affect the choice of gas lift equipment. The cost of the compressor, distribution system for lift gas, and well equipment must be viewed as an economic investment that pays for itself by increasing production. Other forms of artificial lift should also be considered before a final selection is made.

Plunger Lift

Plunger lift is a method of artificial lift used primarily to remove water and condensate from gas wells. It utilizes a plunger, which is a rubber plug, in tubing as shown in Figure 3–11. The cups on the sides of the plunger are flexible and allow fluid to move past as the plunger rests at the bottom of the tubing. On a periodic basis, determined manually or with a timer, the gas valve is opened, and gas is injected below the plunger, lifting it and liquid to the surface. When the plunger reaches the surface, gas injection is stopped, and the plunger falls to the bottom of the well.

Plunger lift is a good method of dewatering gas wells when small amounts of liquid must be removed. Liquid does not slip around the plunger as it would around the top of an injected gas slug.

The plunger can be used to lift a limited column of liquid. The hydrostatic head of several thousand feet of water is so high that considerably more pressure is needed than would be required for gas lift. In the time required for the plunger to reach the surface, the standing valve is held closed, and fluid cannot enter the tubing. Plunger lift cannot be used when high liquid flow rates are anticipated.

Sucker Rod Pumping

Of the more than 500,000 oil wells in the United States, more than 85% are produced using sucker rod lift systems such as the one shown in Figure 3–12. Pumping a well with a sucker rod system requires a surface pumping unit which alternately raises and lowers a steel sucker rod string which, in turn, operates a subsurface positive displacement pump.

Artificial Lift 55

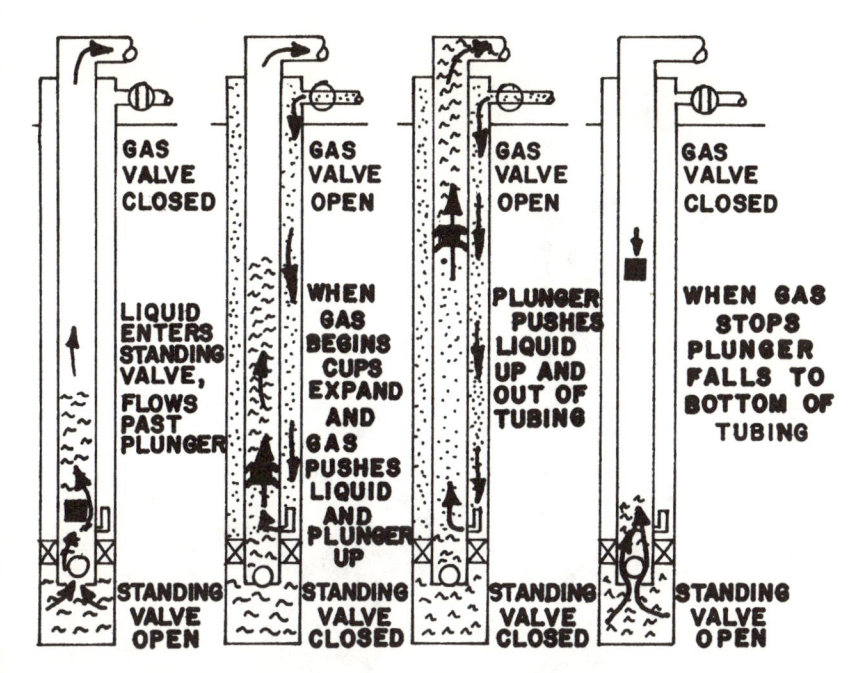

Figure 3-11. Plunger lift uses a rubber plunger (pushed upward with lift gas) to lift fluid from a well.

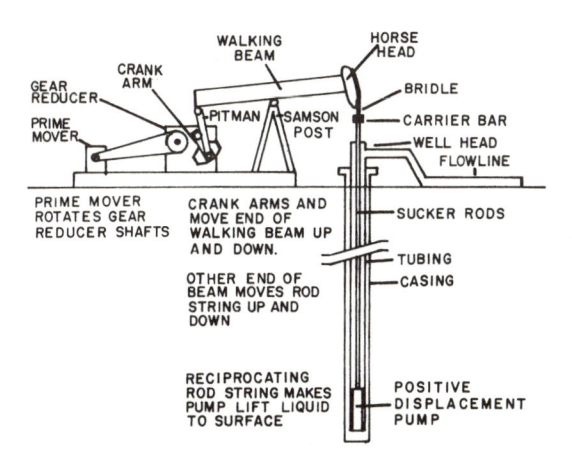

Figure 3-12. Sucker rod pumping systems are mechanical combinations of lever systems, rod strings, and subsurface pumps used to raise fluid from the bottom of a well to the surface.

56 Introduction to Petroleum Production

Figure 3-13. A conventional pumping unit uses the principle of a class I lever system to lift sucker rods.

Surface Pumping Units

Power to raise and lower the rod string is provided by an electric motor or a gas engine. The pumping unit is designed to convert the rotary motion of the drive shaft to the reciprocating motion required by the subsurface pump. The pumping unit is also the principal means of bearing the weight of the rod string and the liquid in the tubing.

Beam Pumping Units. There are several pumping units for sucker rod pumping. Beam pumping units are by far the most common. These pumping units have a large I-beam, called a walking beam, that is used in one of two geometric configurations. The first is a Class I lever system (Figure 3–13) in which the beam is supported at and moves about its center. This is a conventional geometry pumping unit. Figure 3–14 shows the walking beam configured as a Class III lever, and this arrangement is called an improved geometry pumping unit. Another pumping unit (Figure 3–15) is the air-balanced pumping unit, which uses compressed air instead of steel weights for counterbalancing.

The prime mover shaft of a pumping unit rotates at 500–1500 rpm (revolutions per minute). The rod string speed is much slower, 10–20 strokes per

Figure 3-14. An improved-geometry pumping unit uses a class III lever system to operate the rod string.

minute. The reduction in speed is accomplished with sheaves, belts, and a gear reducer.

Figure 3–16 shows how belts deliver power from the prime mover to the pumping unit. The speed at which the unit sheave rotates is related to the speed of the motor sheave by the ratio of the sheave diameters. Although the speed is reduced by the sheaves and belts, the torque, or rotary force, must remain. Several belts on a single set of sheaves may be required to transmit the torque.

V-belts are available in several sizes and lengths. B, C, and D groove belts are frequently used. The letter designates the width of the belt in ascending order.

Figure 3–17 is an illustration of a single reduction gear reducer. Torque is transmitted through the gears while they reduce the speed. Because large shafts are required to transmit large torque, it is sometimes more economical to use double-reduction gear reducers which reduce speed in two stages and use smaller shafts and gears. Figure 3–18 is cutaway view of a double-reduction gear reducer.

One of the limiting factors in the size of a pumping unit is the maximum torque which can be handled by its gears. This is the maximum instantaneous

58 Introduction to Petroleum Production

Figure 3-15. An air-balanced pumping unit uses class III geometry but also uses compressed air for counterbalance.

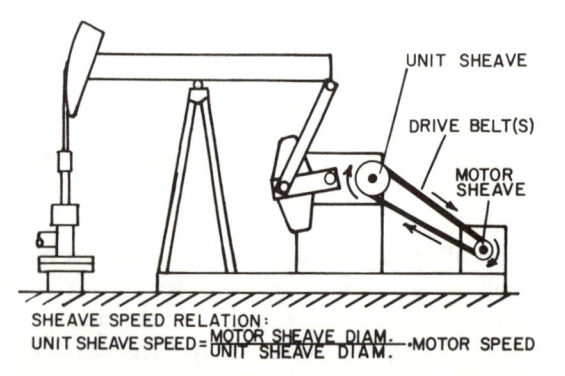

SHEAVE SPEED RELATION:

$$\text{UNIT SHEAVE SPEED} = \frac{\text{MOTOR SHEAVE DIAM.}}{\text{UNIT SHEAVE DIAM.}} \cdot \text{MOTOR SPEED}$$

Figure 3-16. Sheaves, belts, and gear reducers are used to transfer energy from the prime mover to the pumping unit and to reduce rotational speed.

Artificial Lift 59

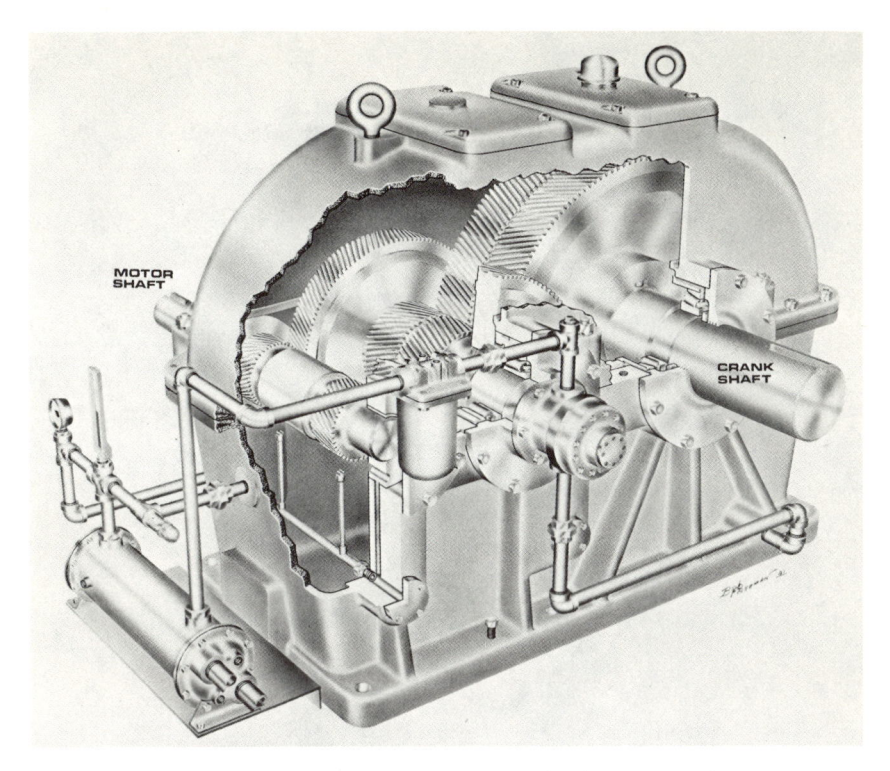

Figure 3-17. Torque is transmitted through a reduction gear reducer on some pumping units. (Courtesy of Lufkin Industries, Inc.)

torque (measured in inch-pounds) that can be handled. One of the API standards rates pumping units in sizes determined by peak torque.

Although it would be possible to raise and lower the rod string simply by the lever system of a pumping unit, this requires great forces from the pumping unit and torques from the gear box and prime mover. Another method is to counterbalance the weight of the rod string with weights (counterweights). This balances the pumping unit much as two people on a seesaw can balance each other. Figure 3–19 shows the effect of counterbalancing in a rod pumping system.

Other Surface Pumping Units

Although beam pumping units are often used, there are other types of pumping units which deserve mention. One unit uses a derrick which has an

60 Introduction to Petroleum Production

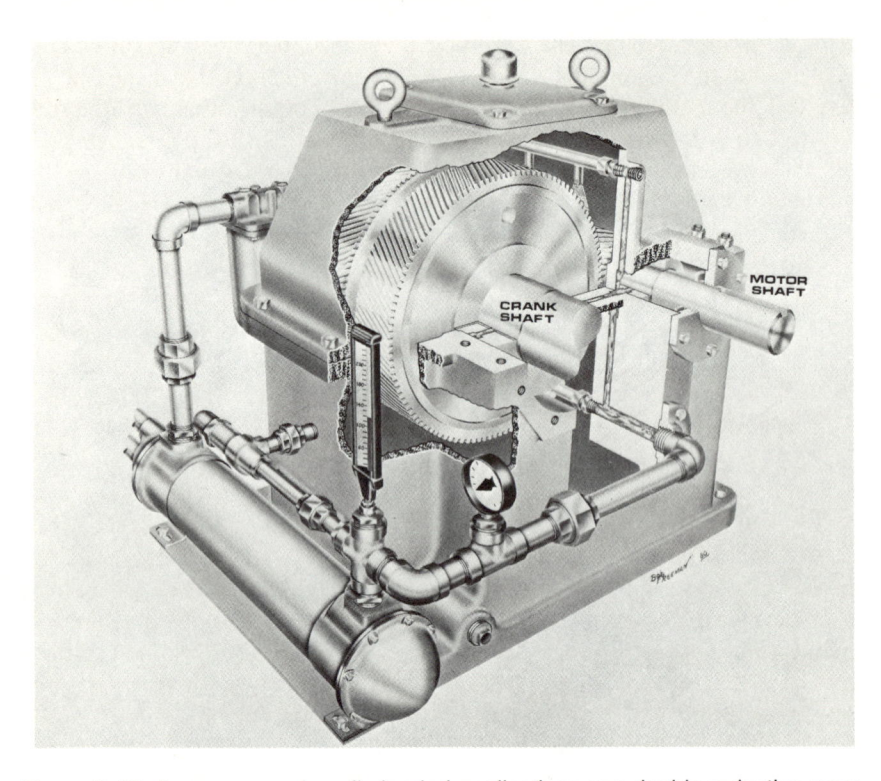

Figure 3-18. Larger, more heavily loaded applications use double-reduction gear reducers. (Courtesy of Lufkin Industries, Inc.)

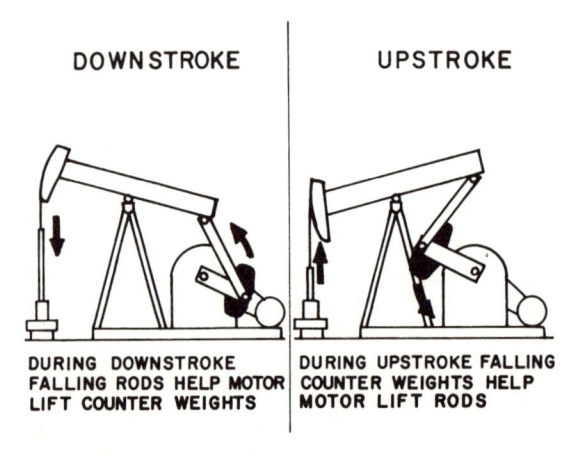

Figure 3-19. Counterbalancing is used to reduce the cyclic loading of the prime mover of a pumping unit.

electric motor in the top. The rod string is attached to a cable that runs across a sheave at the motor and connects to a counterweight that hangs in the derrick. The motor simply rotates the sheave in one direction until the rods are at their lowest position and the weight at its highest. The motor then reverses direction and raises the rods while lowering the weight.

Another sucker rod lift system uses a suspended counterweight much like the one described previously, but the weight is suspended in a hole in the ground. The operation of this system is similar to the derrick unit.

Both of the pumping units described previously enjoyed some popularity for a time; however, their construction and operation were not satisfactory for oilfield use. Both systems have, for the most part, been abandoned.

Prime Movers

The pumping unit is operated by the prime mover—the original power source. This may be a gas engine or an electric motor.

A gas engine used for operating a pumping unit is an internal-combustion engine which uses natural gas as fuel, but gasoline, diesel, or LPG (liquified petroleum gas) may also be used for fuel. The engine may be a four-cycle engine, like automobile engines; or a two-cycle engine, like some outboard boat engines. Low-speed engines (750 rpm or less) or high-speed engines may be used.

Two-cycle, low-speed engines are durable and require no oil filtering and little maintenance, but they are very heavy and expensive. On the other hand, four-cycle, high-speed engines are less expensive, well suited for portable or intermittent service because of light weight, usually have self-contained starters, and can be operated with virtually any fuel, but they require oil filtering and replacement and have a shorter service life.

Many wells which require sucker rod pumping are far from utility power lines, and gas engines are the only way of driving the units. Fuel gas can be natural gas taken directly from the well. However, many sucker rod systems are designed for unattended start-and-stop operation. Gas engines must be started manually, and they are not well suited for unattended operation.

In areas where electric power is available electric motors are preferred for driving pumping units. They are inexpensive, durable, require little or no maintenance, and can be used for unattended or automatic operation.

The electric motors used on pumping units are usually squirrel-cage induction motors that are driven by 480-volt, three-phase electric power. These motors normally run at constant speed; but when a heavy load is exerted on the motors, they slow down slightly without damage. The amount of speed loss the motor can sustain is related to an electrical characteristic of the motor called slip. For pumping unit applications, the maximum possible starting torque and maximum slip are desirable.

62 Introduction to Petroleum Production

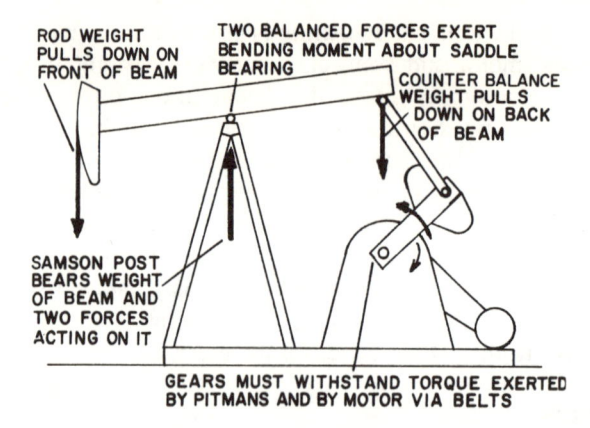

Figure 3-20. Under normal operating conditions, the forces on the samson post are heavy and the difference in forces on the walking beam is small.

The National Electric Manufacturers Association (NEMA) has standard motor designations for selecting slip and starting torque. The NEMA Design D is a motor with slip of more than 5% and starting torque at least 375% of full-load torque. This motor is often used on pumping units.

An electric motor used for pumping is usually left outdoors without any protection from the elements. Two motor housings are used frequently for such service. The totally enclosed, fan-cooled (TEFC) motor is virtually impervious to any weather conditions and is used in particularly harsh climates; however, it is an expensive motor. The open, drip-proof motor is the least expensive outdoor motor and is built to prevent entry of precipitation from an angle not exceeding 15° from vertical.

A new electric motor designed for pumping units is the ultrahigh slip motor (slip exceeding 20%). This motor has gained acceptance over the last few years because it absorbs some of the mechanical shocks by slowing down under heavy load.

Torque and Load on Surface Pumping Units

Torque is a force that causes rotary motion. The torque delivered by an electric motor is transmitted to the pumping unit by the V-belts and the gears and is exerted primarily on the gear reducer. Torque is one of the parameters used in sizing the pumping unit.

The walking beam must be able to bear at least the maximum downward load of the rod string. However, the samson post must bear at least twice this load. Some of the forces exerted on the structure of a pumping unit are shown in Figure 3-20.

Table 3–1
Designations and Ratings of Commonly Used Pumping Units

API Designation	Maximum Torque (inch-pounds)	Maximum Beam Load (pounds)	Maximum Stroke Length (inches)
40–89–42	40,000	8,900	42
57–109–48	57,000	10,900	48
80–133–54	80,000	13,300	54
114–169–64	114,000	16,900	64
160–200–74	160,000	20,000	74
228–246–86	228,000	24,600	86
320–256–100	320,000	25,600	100
456–298–120	456,000	29,800	120
640–356–144	640,000	35,600	144
912–427–168	912,000	42,700	168

(Courtesy of Lufkin Industries Inc.)

The API has a standard nomenclature for pumping units. The API designation gives the maximum torque rating of the gear reducer in thousand inch-pounds, the maximum walking beam load in hundred pounds, and the longest stroke length (the vertical distance from the horse head position at the top of a stroke to the bottom of a stroke) in inches. Table 3–1 is an excerpt from the API standards showing the designation for a few of the more common pumping units.

Installing Surface Pumping Units

Most pumping units are constructed on a unit base made of heavy steel I-beams. The beams and the bracing are made to withstand the stresses exerted by the prime mover, gearbox, and samson post—provided the base is not allowed to move. With the very heavy loads involved in a pumping unit, the base will have the tendency to rock back and forth. Should it be allowed to do so, the effects are the same as bending a paper clip back and forth—the base could break. Once the base integrity is lost, all major components of the pumping unit are subject to stresses which will destroy them.

When a pumping unit is installed, the well site must be prepared to allow the unit base to function properly. The site must be leveled and cleared of vegetation. Many pumping units are installed on pads made of level and compacted clay, sand, caliche (a clay-like material often used for road construction), or gravel as shown in Figure 3–21. The well site must be graded to allow drainage so that water moves away from the unit instead of collecting around it.

Pumping units are mounted on wood or steel beams (sills), or on concrete bases which are placed on the earthen pad. Figure 3–21 shows the use of

64 Introduction to Petroleum Production

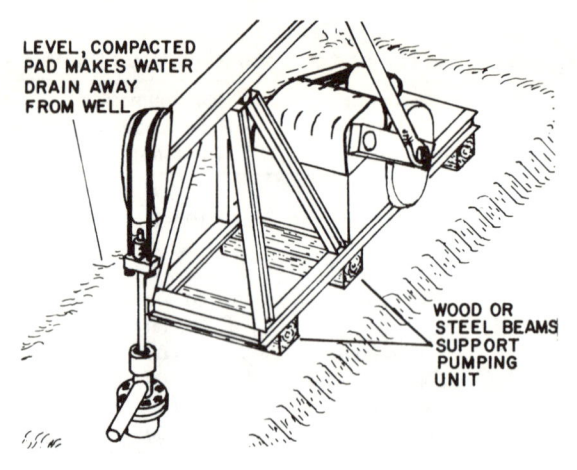

Figure 3-21. The base upon which a pumping unit sits must be carefully prepared to assure proper operation and protection of the equipment.

sills under a unit. Figure 3–22 is a photograph of a preformed, portable unit base. Smaller pumping units (API 228 and smaller) can usually be mounted on sills, but larger units should be installed on concrete bases. The steel base of a pumping unit rests on the beams or cement base and is not allowed to rock if these supports are properly installed.

When a pumping unit is installed, it must be aligned over the wellhead very carefully. If the horse head is not located directly over the wellhead, the rod string can be unduly stressed and can break at the point where it enters the tubing.

Pumping units are made of parts that are bolted together, and these bolts must be kept tight at all times. During normal operation, bolts can loosen slightly; and if they are not tightened, the motion possible at the junction of two parts can cause failure. The base, supports, and all bolts on a pumping unit should be inspected periodically to be sure that no parts of the unit move.

Lubrication and Maintenance

The parts of pumping units that are intended to move against each other are lubricated. Bearings such as those at the pitmens, the cranks, and the joint between the walking beam and the samson post are lubricated with heavy grease. The gearbox, the engine, and some electric motors are lubricated with oil slightly heavier than motor oil. It is imperative that all lubrication points, such as those shown in Figure 3–23, be inspected and refilled periodically.

Artificial Lift 65

Figure 3-22. A portable concrete base may be installed to allow a pumping unit to be firmly anchored.

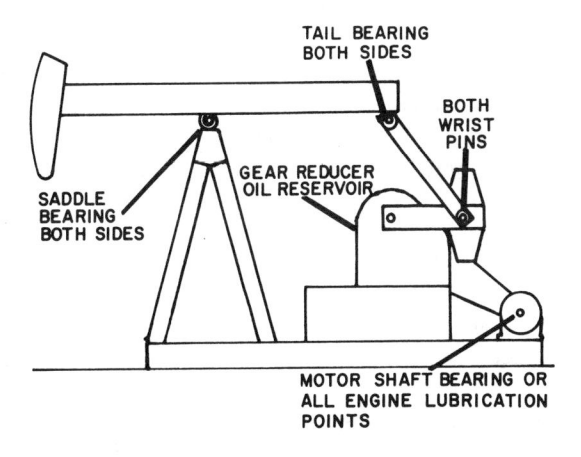

Figure 3-23. Lubrication of all moving parts on a pumping unit is critical to proper operation.

Lack of lubrication is one of the major causes of pumping unit failures. When bearings become dry, the heavy loads cause rapid failure of the metal parts as they rub against each other. Stories abound of pumping units found with the walking beam lying on the ground beside a crumpled samson post because one pitman bearing failed.

66 Introduction to Petroleum Production

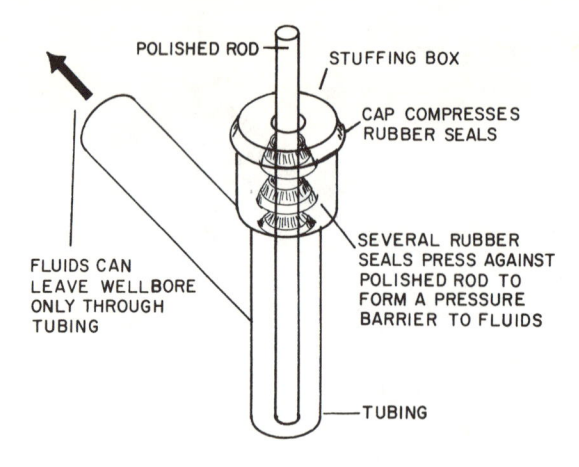

Figure 3-24. A stuffing box forms the seal between pressure inside the tubing and the atmosphere.

Maintenance of pumping units normally amounts to careful periodic inspection. Occasionally, bearings must be replaced, bolts tightened, rust or corrosion removed, and earthwork repacked. The gearbox oil must be checked and changed periodically. Some producers drain gearbox oil, filter and clean it, and reuse the oil.

Rod String

The sucker rod string is a collection of rods, most of which are 25 feet long, coupled to make a single, long, mechanical connection between the surface pumping unit and the subsurface pump located near the bottom of the tubing. Although a single rod seems inflexible, the rod string, which may be several thousand feet long, is a dynamic system with kinetic characterisitics that affect not only the operation of the pumping unit and the pump but also the string's behavior and lifetime.

Polished Rod and Stuffing Box. The topmost element of the rod string is the polished rod. This rod has a diameter of 1.25-1.50 inches and an outside surface of polished chrome or stainless steel. It is against this very smooth surface that the stuffing box must act to form a seal between the moderate pressure in the tubing and the atmosphere to keep liquids and gases inside the tubing. Figure 3-24 shows the arrangement of the polished rod and the stuffing box.

A stuffing box which houses rubber cups called stuffing box rubbers (Figure 3-25) is a mechanical fitting on the top of the top tubing joint. When the top of the stuffing box is unbolted and raised, old rubbers may be removed and new ones slipped around the polished rod and pushed into place. The top

Artificial Lift 67

Figure 3-25. The stuffing box is the principal seal around the polished rod between the moderate pressure in the tubing and low atmospheric pressure. (Courtesy of Midland College.)

of the stuffing box compresses these cups and presses them against the polished rod, forming a fluid-tight pressure seal even when the rod is moving.

Because polished rods are made of thick metal, they are expensive. A less expensive alternative is to use a polished rod liner (Figure 3–26). A polished rod liner is made of chrome and is attached to the top rod in the rod string. The stuffing box rubbers form a seal against the liner. The top rod with the liner is much less expensive than a polished rod.

Sucker Rods. These rods are shaped as pictured in Figure 3–27 and have the various parts shown. The overall length of a rod is 25 feet, but shorter rods (pony rods) are available in lengths as short at two feet.

Rods are made of steel alloys, the components of which determine hardness, brittle character, and susceptibility to corrosion. Carbon steel is an alloy of iron, carbon, manganese, phosphorus, sulfur, and silicon. Other elements, such as nickel, chromium, and molybdenum, are added to the steel used in sucker rods. The rods are cast into ingots, rolled and forged into the proper shape, heat tempered, and finally machined into the exact shape so that sucker rods made by several manufacturers are interchangeable.

68 Introduction to Petroleum Production

Figure 3-26. Polished rod liners offer an alternative to an expensive solid polished rod.

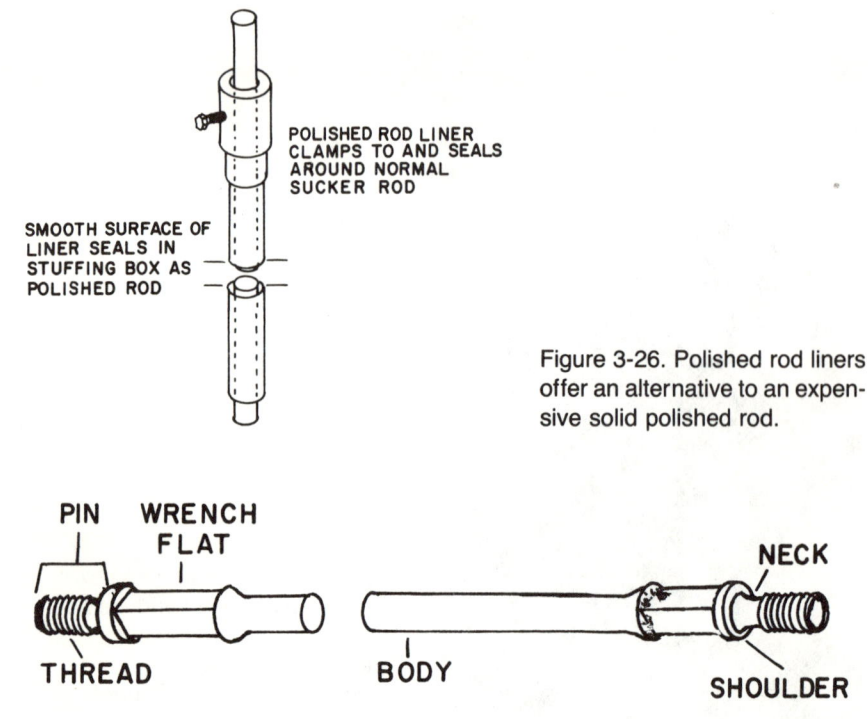

Figure 3-27. Sucker rods form the connection from the surface pumping unit to the subsurface pump.

The API has established standards for sucker rod alloys, dimensions, strengths, methods of handling, and calculations for proper sizing. Table 3–2 shows the standard API grades, the average compositions of the alloys, and the normal applications for the rods.

Like other steels, the metal used for sucker rods has rated yield strength. Because sucker rods are used with very heavy, reversing loads, it is not advisable for rods to be used with loads even approaching the yield of the metal. The API, in establishing standard rod grades, also established the maximum allowable stress (load in pounds per unit of rod cross-sectional area in square inches/psi) for each rod grade.

After steel is alloyed and heat treated, the metal molecules align in a crystalline structure in which there are spaces between the metal grains. Molecules of hydrogen, the smallest molecules in nature, are smaller than the spaces between grains of some steel alloys. Molecular hydrogen, present in quantity in hydrogen sulfide, can invade the crystal structure and weaken it by making the alloy very brittle in a phenomenon called hydrogen embrittlement. When hydrogen sulfide is an appreciable component of the produced

Artificial Lift   69

Table 3–2
Characteristics of Standard Sucker Rods

API Grade	Principal Alloy Components	Maximum Allowable Stress (psi)	Service
C	Carbon Manganese	28,000	Non-corrosive Average load
D	Nickel Chrome Molybdenum	40,300	Heavy loads Deep wells
K	Nickel Molybdenum	31,400	Hydrogen Sulfide

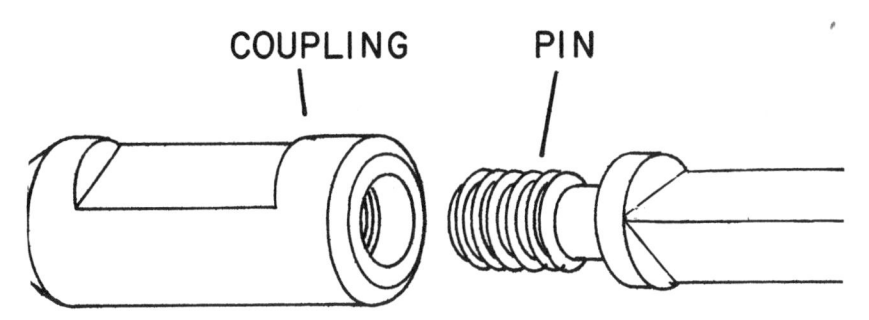

Figure 3-28. The pin and coupling used to join two sucker rods are designed to operate together to form a strong joint.

gases, the K-type rod should be used because its crystal structure is not affected by hydrogen as are some other rods.

Corrosion by water and some chemical compounds as well as hydrogen embrittlement are very serious problems in rod strings. Some rods are treated by a heating process called *annealing*, which creates a microscopically thin, very hard surface resistant to corrosive environments. As long as this surface is intact, the rod is less likely to be attacked by corrosion. Another process which also creates a very hard surface on rods is *shot peening* (a process similar to that used by blacksmiths to harden metal) in which tiny steel pellets (shot) are propelled against the rod by compressed air. The continued battering by the equivalent of tiny hammers creates a hard surface on the rod without changing the shape or apparent finish of a rod.

Rods are joined with a rod coupling or box as shown in Figure 3–28. The pin and coupling threads are tapered; when they are tightened, they stretch the pin neck slightly. The neck, in trying to recoil, pulls the pin shoulder tightly against the coupling. To prevent stagnant liquid from collecting and

Table 3-3
Standard Sizes and Strengths of Sucker Rods

Rod Diameter (inches)	Coupling Outside Diameter (inches)	Rod Weight (lbs/ft)
0.500	1.000	0.72
0.625	1.250	1.15
0.750	1.500	1.64
0.875	1.813	2.20
1.000	2.000	2.88
1.125	2.375	3.68

causing corrosion inside the coupling and to keep the junction from unscrewing, the pin and coupling must be properly tightened. Both are equipped with flat areas called *wrench flats* so that a smooth wrench can be used instead of a pipe wrench with teeth that will cut the hardened surface of the rod and coupling.

Rod couplings are normally the weakest part of a rod string. The couplings must be made of steel comparable in strength to that of the rods. The API has issued standards for rod couplings that reflect the alloy used to make the couplings. These types are:

1. *Class T* (high strength, resistant to hydrogen, embrittlement, moderately resistant to corrosion)
2. *Class S* (used for heavy loading, mildly resistant to corrosion)

Couplings are annealed or shot peened for corrosion protection just as are rods. A corrosion-resistant coupling that has been enjoying some recent success is the sprayed-metal coupling. In a process using electric arcs through metal tips metal such as high-chrome steel is melted and blown onto a normal coupling by a jet of compressed gas. This spray of molten, corrosion-resistant metal forms a thin metal coating on the coupling. This coating has proved to be more resistant to corrosion and hydrogen embrittlement than many other models.

Rods and couplings are made in the standard sizes shown in Table 3-3. The pins and couplings are manufactured according to standards so that they may be interchanged at any time. Couplings are also manufactured so that two different rod sizes may be joined. Slim hole couplings that are thin-walled versions of standard couplings are also available when space is at a premium.

Rod Stretch. The metal in rods stretches and rebounds much like spring steel. When the weight of the rod string and the weight of all fluid in the tubing is placed on the rods, they stretch several inches. An example of rod stretch is a weight suspended from a rubber band:

> If a fishing weight is suspended from a rubber band, the band stretches slightly. If more weights are added, the rubber band stretches more—until its yield point is reached, and the rubber band breaks. However, a larger rubber band could be substituted to support the weight that broke the smaller band.

The analogy of a rubber band to sucker rods is an almost perfect one, except for the order of magnitude. If a rod has load added, it will stretch slightly. As more weight is added, the rod stretches more. Before the rod breaks under load, it could be replaced with a larger rod, and the weight increase continued without harm.

Anything that decreases the strength of a rod decreases the amount of weight it can support. By analogy, cutting through half the width of a rubber band reduces the amount of weight it can support; corrosion and hydrogen embrittlement reduce the strength of part of the cross section of a rod and effectively reduce the size of the rod. Undue stress on the rod, such as bending or striking, has much the same effect as fatiguing a paper clip by repeatedly bending it.

Rod Care. The sucker rod string is a dynamic system in which operation must be precise and controlled. Rods cannot perform properly for extended periods unless they are properly cared for. Proper rod care includes handling, storage, and installation procedures.

When rods are handled, care must be given to keeping the rods straight and seeing to it that nothing scratches the hard surface of the steel. New rods are shipped in bundles that use wood spacers and steel straps to hold the rods rigidly in place. Lifting chains or cables should be positioned to keep the bundles straight. New or used rods should be stacked on wood sills positioned to support each end and the middle to keep the rods from bending.

When rods are stacked, nothing should be allowed to fall on them because falling objects can scratch the hard outside surface. To avoid bending or scratching rods, workers should never walk or jump on them.

Any lifting or moving force that causes sharp bends in rods must be avoided. When a rod is bent, one side is stretched while the other is compressed, as indicated in Figure 3–29. Sharp bends can permanently stress the steel and cause a weak place in the rod.

Rods are manipulated by a well-servicing unit with elevators, which are lifting devices sized to support rods just below the wrench flats. If an elevator

72 Introduction to Petroleum Production

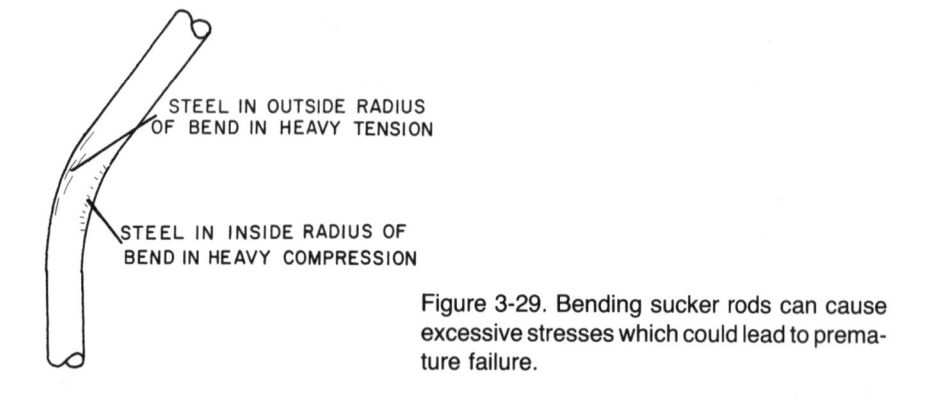

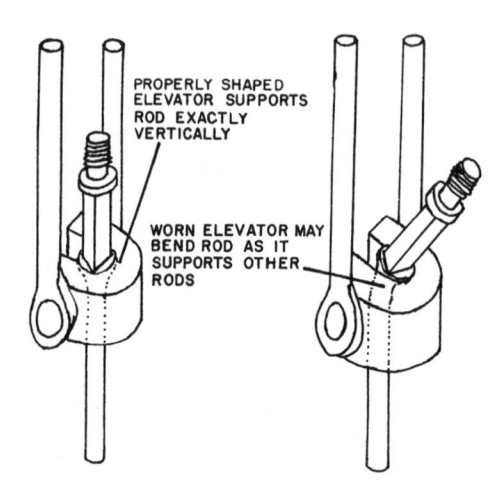

Figure 3-29. Bending sucker rods can cause excessive stresses which could lead to premature failure.

Figure 3-30. Worn elevators can bend sucker rods near the wrench flats.

(Figure 3–30) is worn slightly, it can bend the rod as the weight of the rod string is supported by the rod. Elevators should be inspected frequently to assure that they are not worn.

Any action that could scratch or cut the outside surface of sucker rods must be prevented. The hard, outer surface is very resistant to most forms of corrosion. However, any scratch or cut in this surface not only exposes the softer, inside metal to corrosive activity, but also concentrates and accentuates corrosion in this spot.

For the pin-coupling junction to perform properly, the pin neck must be stretched slightly. If the pin is stretched too much or not enough, as shown in Figure 3–31, the neck can be overstressed or can be forced to flex too much while allowing fluid to enter the joint. The pin neck is stretched by applying torque to the pin and coupling. Applying torque that stretches the neck ex-

Artificial Lift 73

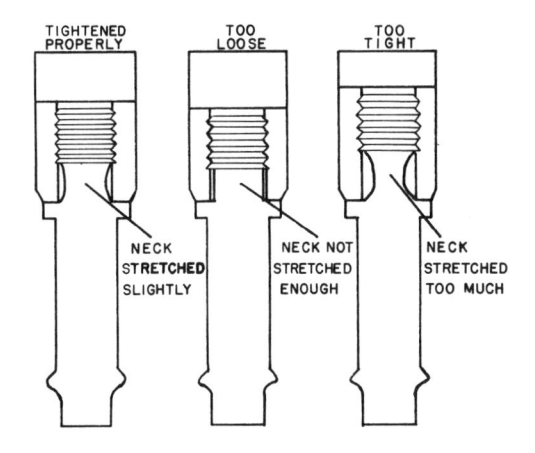

Figure 3-31. The pin is designed to be stretched slightly to provide a strong joint with a coupling.

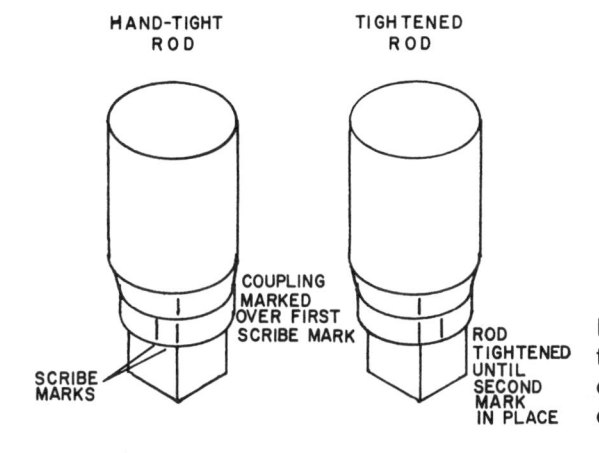

Figure 3-32. Circumferential displacement assures each joint is properly tightened.

actly, called proper rod makeup, is critical. The API has issued a standard method of rod makeup, called the circumferential displacement method, which ensures that the correct torque is applied to every joint. Two marks are scribed in the shoulder of every rod manufactured by API specifications, and the distance between these marks varies with rod size. When the coupling of one joint is threaded into the pin of another until the shoulder meets the coupling "hand tight," a mark is placed on the coupling as shown in Figure 3–32, and the joint is tightened until the second shoulder mark meets the coupling mark. When this method is properly employed, every joint is threaded precisely. If power tongs are used, they must be recalibrated frequently to assure proper displacment.

Proper rod makeup depends on the pin and coupling threads being completely clean and well lubricated. When rods are installed or when they are

74   Introduction to Petroleum Production

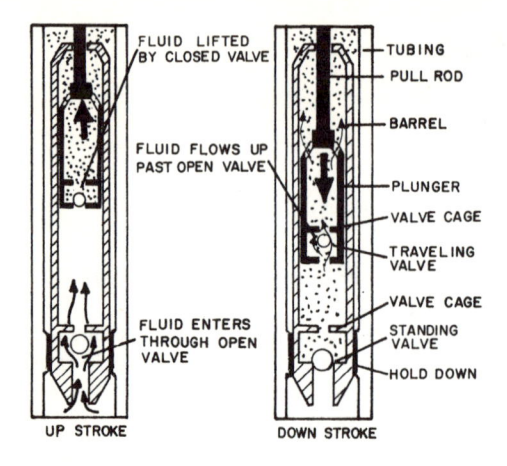

Figure 3-33. The standing valve and barrel remain stationary in a stationary barrel pump while the plunger and traveling valve move.

pulled and placed back in a well during servicing, the pins and couplings should be thoroughly cleaned with a course brush and detergent and lubricated. One successful method of both cleaning and lubricating is to scrub the pins and couplings with a mixture of detergent, light lubricating oil, and corrosion inhibitor. The corrosion inhibitor gives future protection against corrosion caused by well fluids entering the junction.

Non-Ferrous and Coated Rods. In an effort to avoid some corrosion problems which can be encountered in rod pumped systems, sucker rods made of materials other than steel have been developed. The most notable of these materials is virtually indestructable fiberglass. For the pin-coupling joints to operate properly, these parts are made of steel and are bound to the fiberglass body with epoxy adhesive.

Another method used for corrosion protection is plastic coating the rods. This plastic is applied to, and adheres to, standard rods. This method is usually only moderately successful because the plastic is easily scraped off as the rods rub against tubing.

Subsurface Pumps

The pumping unit and rod string are necessary for the operation of this lift system, but the heart of a sucker rod pumping system is the subsurface pump itself. The basic operating parts of a pump are shown in Figure 3–33. This configuration is called a *stationary barrel pump* because the barrel remains fixed while the plunger moves inside it. As the rods pull the plunger up, the hydrostatic head of the tubing fluid pushes the traveling valve (a ball and seat valve like the standing valve discussed earlier) closed. As the plunger continues upward, the pressure between valves is low, and bottom hole pressure

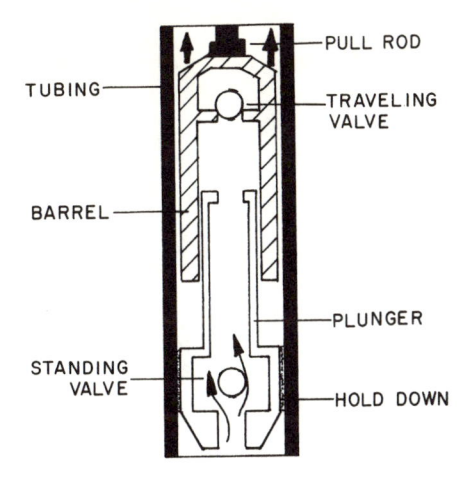

Figure 3-34. The plunger and standing valve remain stationary in a traveling barrel pump while the barrel and traveling valve move.

opens the standing valve and pushes liquid into the barrel. As the rods begin to move downward, the standing valve closes immediately. With continued downward motion, the pressure between valves increases until this pressure exceeds the tubing hydrostatic head and opens the traveling valve, allowing the liquid between valves to move above the traveling valve. This portion of liquid is lifted as the next upstroke begins.

When the rod string moves upward, the traveling valve is closed, and the hydrostatic head of the tubing liquid is exerted against the valve. Thus, the rods bear their weight and that of the tubing liquid while the tubing bears only its own weight. On the downstroke, the traveling valve opens, and the rods bear only their own weight while the tubing bears its weight and that of the tubing liquid.

The plunger and barrel must form a tight liquid seal so that liquid is lifted by the closed traveling valve instead of leaking by the plunger. This seal is formed by using precisely machined metal parts in both the plunger and barrel. The machined parts fit together so closely (a few thousandths of an inch) that little, if any, liquid leaks by the plunger. Any minute fluid leak acts as lubrication for the metal parts. The plunger for this sealing method is called simply a metal plunger or plain-metal plunger. A grooved metal plunger has grooves cut into the plunger which assist in the fluid seal. Another method for sealing against liquid flow past the plunger is to use fiber rings which act much like washers. This is called a soft-packed plunger.

Figure 3–34 shows another subsurface pump called a *traveling barrel pump*. This pump's operation is similar to the operation of a stationary barrel pump, except that the traveling valve is attached to the barrel.

76 Introduction to Petroleum Production

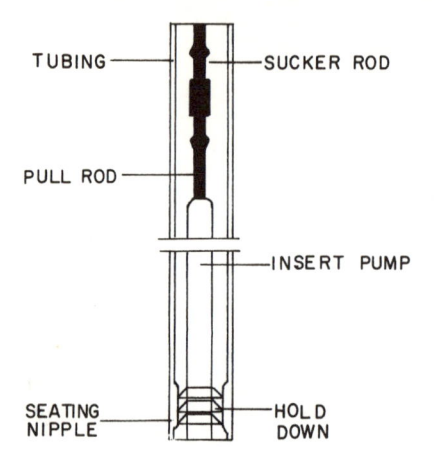

Figure 3-35. All parts of an insert pump are contained in one housing and are attached to the end of the rod string.

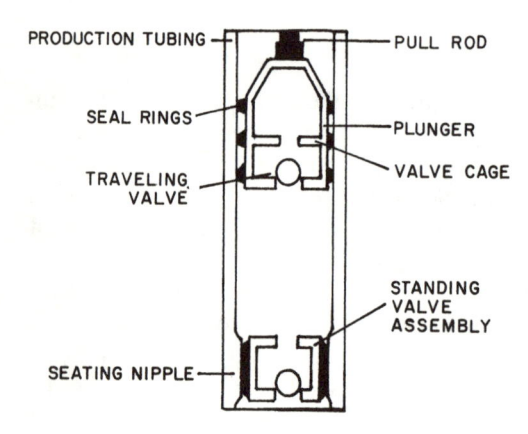

Figure 3-36. The barrel and standing valve of a tubing pump are attached to the pump while only the plunger and traveling valve are attached to the rod string.

The pump and tubing work together as a system. The pump may be a complete unit which is attached to the rod string. The pump is lowered into the tubing and attached to a seating nipple by plastic, fiber, or mechanical cups called *hold-downs*. This pump (Figure 3–35) is called an insert pump or rod pump. In some pumping systems one of the bottom tubing joints is the pump barrel. The standing valve, attached to the traveling valve for installation and detached for operation, is lowered and mounted in the seating nipple. Such a pump is called a tubing pump and is shown in Figure 3–36.

Comparison of Subsurface Pump Types. Each pump type has applications for which it is well suited. Insert pumps are often used in wells where solids are produced, where pumps have to be changed often, when secondary

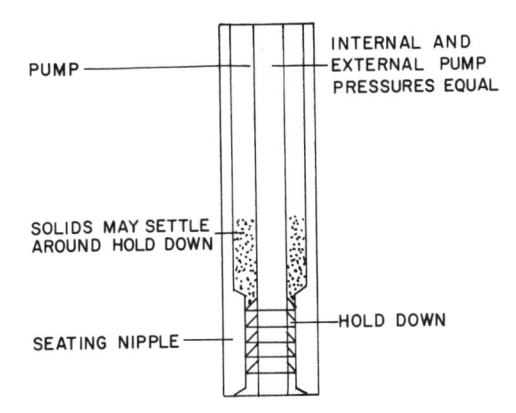

Figure 3-37. Solids may collect around a pump using bottom hold-downs, but the barrel may be made of thin metal.

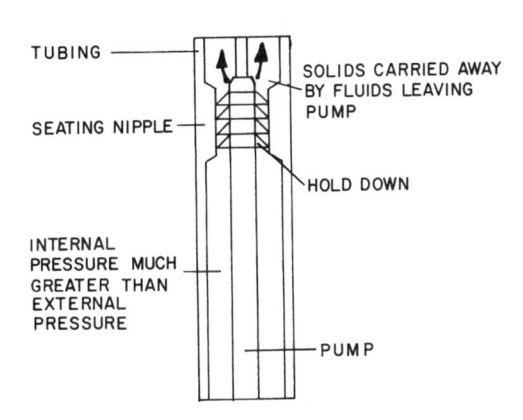

Figure 3-38. Solids are not likely to collect around a pump using top hold-downs, but the barrel must be made of thick metal.

recovery operations cause variations in production rates, and any other times when the pump must be removed frequently. Insert pumps can be removed easily when damaged or when a pump size change is required. Also, insert pumps are used in deep wells where the plunger-barrel seal is made difficult by the great hydrostatic pressure and metal plungers are required.

In deep wells the location of the hold-down can be critical. If the hold-down is at the bottom of the insert pump, the hydrostatic head of tubing liquid is exerted on both the inside and outside of the barrel, as shown in Figure 3-37, and a thin-walled barrel is acceptable. However, any solids pumped with liquids will settle around the pump and possibly wedge it in the tubing. On the other hand, a top hold-down pump (Figure 3-38) can be used to wash away solids continuously, but a much heavier barrel is required to contain the great difference in pressure between inside and outside without bursting. A compromise between these two possibilities is the use of both top and bottom hold-downs, which combine the advantage of both but require a special seating-nipple arrangement.

Traveling barrel pumps are often used when a great deal of sediment is pumped with liquids. The continuous agitation lessens the possibility of wedging the pump in the tubing, but there is a danger of excessive wear when the barrel rubs against the tubing.

Insert pumps are efficient pumps in terms of the liquids lost to leakage by the plunger barrel seal. The metal plunger pump is the most efficient of all, but it is also the one most susceptible to damage by solids scratching the barrel and plunger and ruining the close tolerances. Insert pumps are usually limited in volume because it is very difficult to maintain the close tolerances required in large pumps.

Tubing pumps are used for high-volume, shallow wells. Because the plunger is lowered into the barrel, it is not possible to attain the close tolerances required for high efficiencies; thus, deep pumping is not possible because of the great hydrostatic pressure. However, because all the parts of the tubing pump are much larger than for most insert pumps, it can handle greater volumes.

In another of its standards the API has assigned a naming method that assures that the correct pump is obtained regardless of the manufacturer used. Table 3–4 shows the API standard nomenclature for subsurface pumps. When the correct codes have been inserted in the API number, the only facts about a pump left unspecified are materials of construction and valve sizes.

Some pumping applications require specialized pumps that are not described by the API standard. One such pump is the so-called "three-tube" pump shown in Figure 3–39. This pump uses liquid seals with metal valves to lift liquid containing significant volumes of solids, such as sand, used in a fracture stimulation. This pump is not easily damaged by solids and can be used to clean a wellbore after stimulation, but this pump is very inefficient and must be replaced with another more efficient pump as soon as the solids have been removed from a well.

Subsurface Pump Displacement. Subsurface pumps are called positive displacement pumps because on each stroke a fixed volume of liquid is lifted. Figure 3–40 illustrates the volume lifted on each stroke. This volume is related to the cross-sectional area of the pump and the length of the stroke. For a given pump, the rate at which liquid is pumped is determined by the stroke speed (the number of strokes per minute). If the stroke speed is increased, the pump rate is increased as shown in Figure 3–40.

Table 3-4
Pump Designation

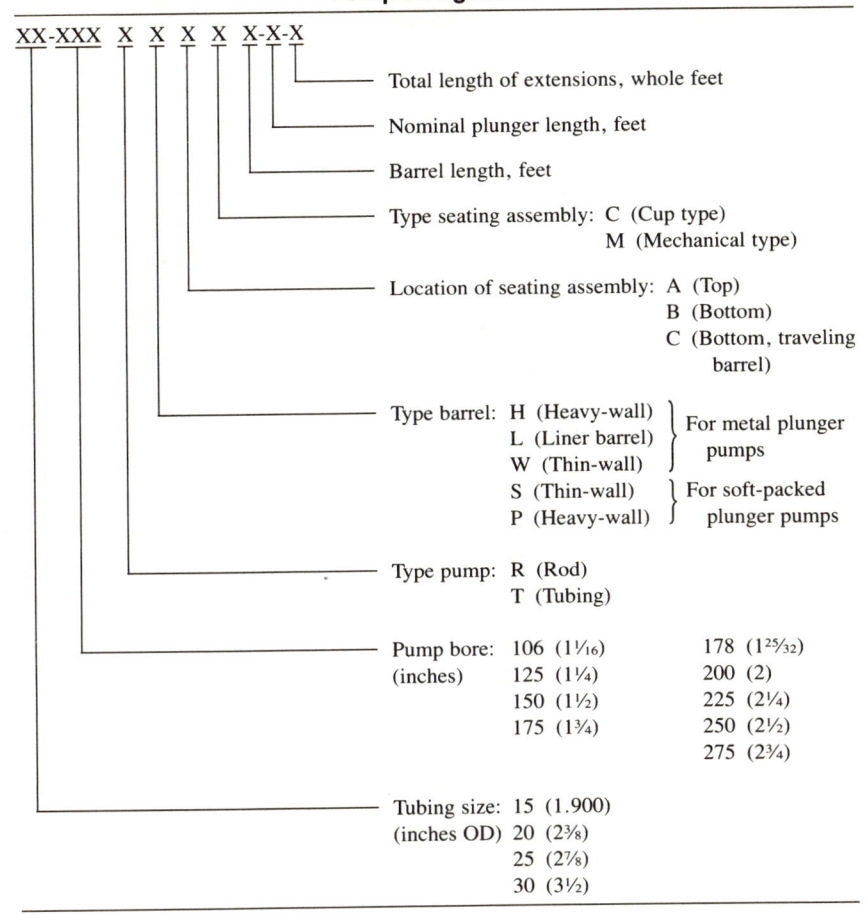

```
XX-XXX X X X X X-X-X
                │ │ │
                │ │ └── Total length of extensions, whole feet
                │ └──── Nominal plunger length, feet
                └────── Barrel length, feet
```

- Type seating assembly: C (Cup type)
  M (Mechanical type)
- Location of seating assembly: A (Top)
  B (Bottom)
  C (Bottom, traveling barrel)
- Type barrel: H (Heavy-wall) ⎫
  L (Liner barrel) ⎬ For metal plunger pumps
  W (Thin-wall) ⎭
  S (Thin-wall) ⎫ For soft-packed
  P (Heavy-wall) ⎭ plunger pumps
- Type pump: R (Rod)
  T (Tubing)
- Pump bore: 106 (1 1/16)   178 (1 25/32)
  (inches) 125 (1 1/4)   200 (2)
            150 (1 1/2)   225 (2 1/4)
            175 (1 3/4)   250 (2 1/2)
                           275 (2 3/4)
- Tubing size: 15 (1.900)
  (inches OD) 20 (2 3/8)
              25 (2 7/8)
              30 (3 1/2)

Letter Designation

	Metal Plunger Pumps			Soft-Packed Plunger Pumps	
Type of Pump	Heavy-Wall Barrel	Liner Barrel	Thin-Wall Barrel	Heavy-Wall Barrel	Thin-Wall Barrel
Rod Pumps					
Stationary barrel, top anchor	RHA	RLA	RWA		RSA
Stationary barrel, bottom anchor	RHB	RLB	RWB		RSB
Traveling barrel, bottom anchor	RHT	RLT	RWT		RST
Tubing Pumps	TH	TL		TP	

(Courtesy of the American Petroleum Institute)

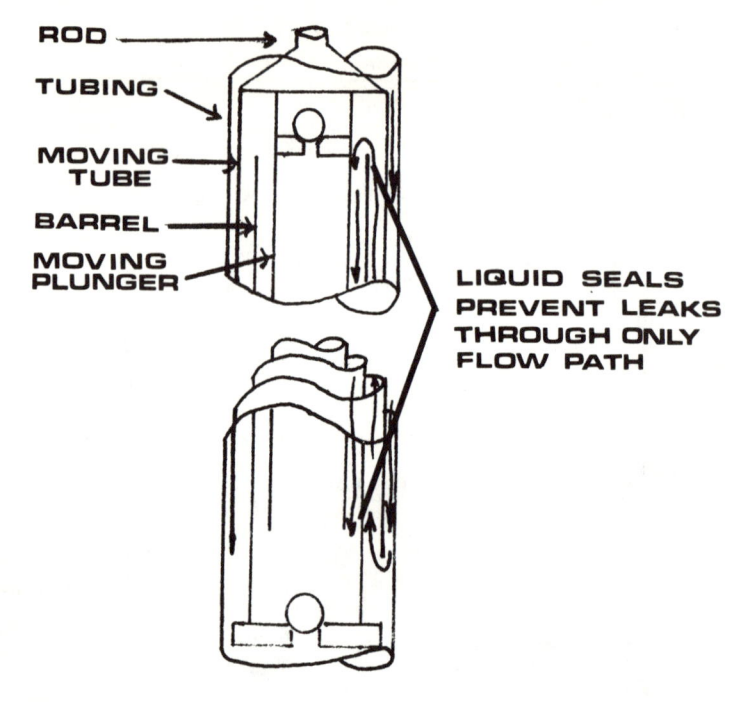

Figure 3-39. In a three-tube pump the liquid between the tubes prevents fluid flow without using metal-to-metal or soft-packed seals.

Tubing and Subsurface Equipment

The tubing used for sucker rod pumping is the same as that discussed earlier. On each stroke of the pump, the weight of tubing liquid is transferred from the rod string to the tubing and back. The weight of tubing liquid is several thousand pounds, and the transfer of this much weight can make the tubing string stretch several feet and rebound on each stroke, as shown in Figure 3–41. This stretching action may eventually overstress the tubing, but the increase in tubing length changes the pump stroke length and can affect the pumping rate.

A tubing anchor, pictured in Figure 3–42, is used to couple the tubing to the casing with slips. A tubing anchor allows fluid to move in the annulus. When an anchor is engaged with the casing and the tubing left in tension, the weight of tubing liquid is transferred through the anchor to the casing; any potential movement is prevented.

The seating nipple used for sucker rod pumping is the same as discussed earlier. The hold-downs of the pump are placed in the seating nipple to couple the pump and tubing.

Artificial Lift 81

ON EACH COMPLETE STROKE, THE PUMP DISPLACES THE VOLUME OF A CYLINDER WHOSE DIAMETER IS THE INSIDE OF THE PLUNGER AND WHOSE HEIGHT IS THE STROKE LENGTH.

INSIDE DIAMETER

STROKE LENGTH

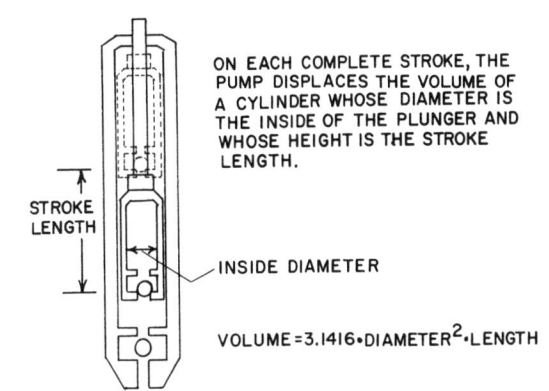

$$\text{VOLUME} = 3.1416 \cdot \text{DIAMETER}^2 \cdot \text{LENGTH}$$

Figure 3-40. During each stroke, a subsurface pump lifts a specific volume of liquid.

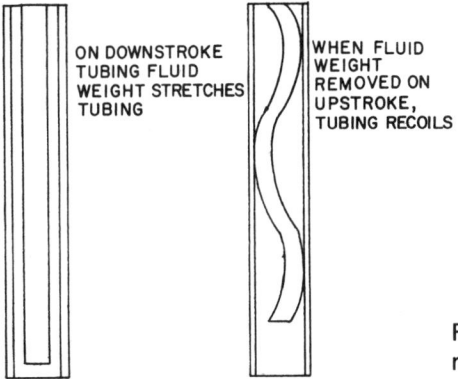

ON DOWNSTROKE TUBING FLUID WEIGHT STRETCHES TUBING

WHEN FLUID WEIGHT REMOVED ON UPSTROKE, TUBING RECOILS

DOWNSTROKE   UPSTROKE

Figure 3-41. Tubing can stretch and rebound during each stroke unless it is anchored to the casing.

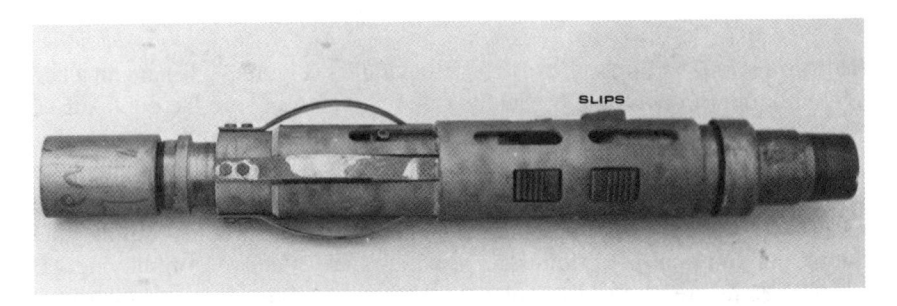

SLIPS

Figure 3-42. A tubing anchor is attached to tubing and prevents tubing movement in the casing.

82 Introduction to Petroleum Production

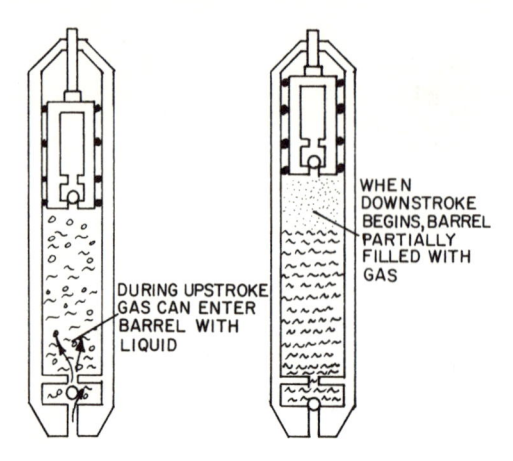

Figure 3-43. Gas interferes with the efficiency of a subsurface pump.

Lift Efficiency and Gas Interference

Subsurface pumps are designed to pump incompressible liquid. When gas enters a pump, as shown in Figure 3–43, the gas takes space that could be occupied by liquid. The pump displacement capacity for lifting pure liquid is then reduced. Pump efficiency (the ratio of actual pump displacement to pump displacement under ideal conditions of lifting liquid with valves that do not leak) is reduced when gas enters the pump. Pumping compressible liquid (oil with a significant volume of dissolved gas) can also reduce pump efficiency because saturated oil occupies more volume than oil with no solution gas. Finally, pump efficiency is reduced when one or both valves, particularly the traveling valve, leak as a result of mechanical wear or sedimentation. On the upstroke, hydrostatic pressure can cause tubing liquid to flow past the traveling valve, thereby losing part of the pump displacement. On the downstroke, the tubing hydrostatic head is exerted against the standing valve and can push fluid past this valve with the same result.

Bottom Hole Gas Separators. These separators (Figure 3–44) are attached to the bottom of the tubing below the standing valve and are the only path of entry into the pump. Liquids and gas are forced to flow through the torturous channels of the separator, and the many direction changes of the fluid cause solution and free gas to separate from the liquid. The gas, following the path of least resistance, exits the separator in the annulus, while the liquids enter the tubing and pump. Bottom hole separators are effective in removing gas before liquids are allowed to enter the pump. However, formation particles of sand, scale, or paraffin can easily enter the small flow channels in the separator.

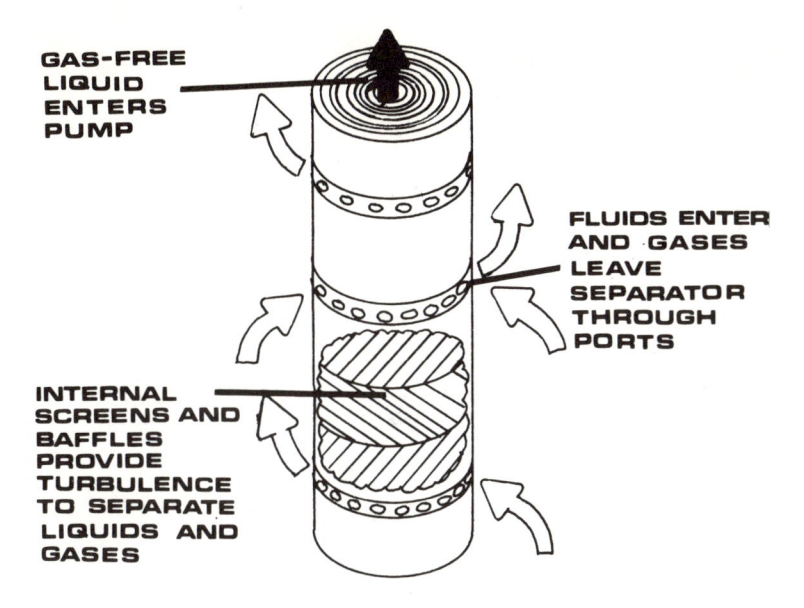

Figure 3-44. A bottom hole separator attached to the bottom of a pump separates liquids and gases and prevents gases from entering the pump.

Gas and Mud Anchors. Figure 3-45 shows a bottom hole arrangement which is also effective in separating liquid from gas. The mud anchor is one joint or more of large tubing which is plugged at the bottom and perforated near the top. The mud anchor is attached to the seating nipple and installed with the tubing. The gas anchor is smaller tubing attached to the bottom of the insert pump. The size and number of perforations and the relative sizes of the two anchors are critical in optimizing adequate gas separation while maintaining unrestricted fluid flow. Gas separation is maximized by causing restriction to flow, but fluid flow is impeded by such restriction. Any solids that are carried by liquid usually settle in the mud anchor and can eventually clog it.

Sucker Rod Pump System Operation

As mentioned earlier, the sucker rod string operates much like a rubber band with a small weight attached to its end. The rods are moved up and down, and the rod string stretches during the upstroke and rebounds during the downstroke. Thus, the stroke length at the pump usually differs from the surface stroke length by several inches. On each upstroke, the traveling valve closes, places the full tubing fluid weight on the rods, and causes additional

84 Introduction to Petroleum Production

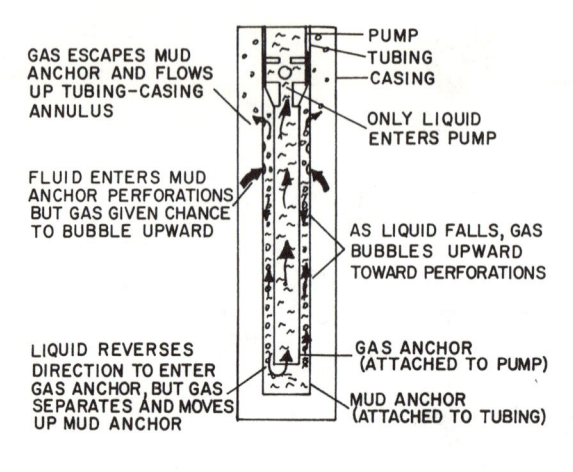

Figure 3-45. A mud anchor and gas anchor can be used to keep gas from entering a subsurface pump.

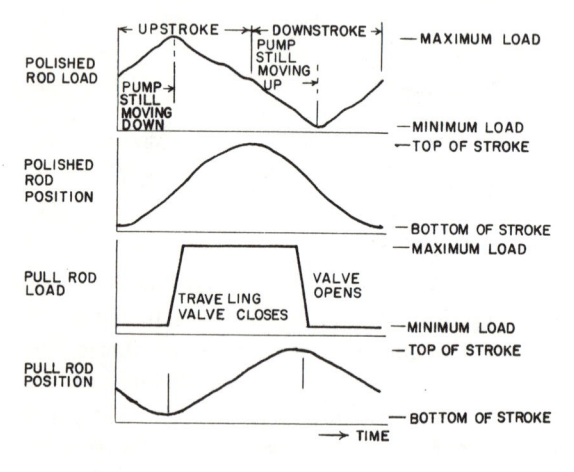

Figure 3-46. Polished and pull rod load and position can be graphed versus time to show their variation.

rod stretch. On the downstroke, when the rods are rebounding naturally, the traveling valve opens, removes some of the rod load, and causes the rods to rebound more than expected without a change in rod load.

A graph of the surface rod load and rod load at the pump versus time is shown in Figure 3–46. The graph shows the effect of rod stretch on polished rod load, the surface rod. The position of the polished rod and pump position versus time are also shown in Figure 3–46. It is not at all unusual for the pump to be moving in one direction, while the polished rod moves in the other direction because of rod stretch.

Figure 3–47 is a different presentation of the same information of Figure 3–46. In this case the polished rod load is plotted versus polished rod position, and the corresponding pull rod load is plotted versus the pull rod position.

Artificial Lift

[Figure showing dynagraph diagrams]

POLISHED ROD LOAD VERSUS ROD POSITION—DYNAGRAPH

- BOTTOM OF STROKE
- UPSTROKE
- TOP OF STROKE
- MAXIMUM POLISHED ROD LOAD
- RODS BEGIN TO RELAX AFTER ROD STRING REACHES MAXIMUM VELOCITY UP
- DOWNSTROKE
- MINIMUM POLISHED ROD LOAD
- RODS STILL STRETCHING AS TRAVELING VALVE CLOSES AT START OF UPSTROKE
- RODS BEGIN TO STRETCH AFTER ROD STRING REACHES MAXIMUM VELOCITY DOWN
- RODS STILL RELAXING AS TRAVELING VALVE OPENS AT START OF DOWNSTROKE

PULL ROD LOAD VERSUS ROD POSITION

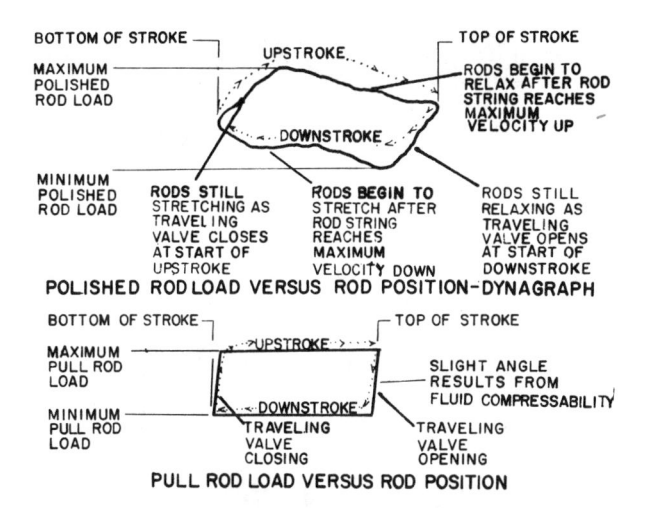

- BOTTOM OF STROKE
- UPSTROKE
- TOP OF STROKE
- MAXIMUM PULL ROD LOAD
- SLIGHT ANGLE RESULTS FROM FLUID COMPRESSABILITY
- DOWNSTROKE
- MINIMUM PULL ROD LOAD
- TRAVELING VALVE CLOSING
- TRAVELING VALVE OPENING

Figure 3-47. Rod load can be graphed versus rod position to indicate loading and total work done by the system.

Figures 3-46 and 3-47 show that there is significant stretch and relaxation in the rod string. Although it appears that the rods are compressed as they rebound, this is not the case. The rods are stretched even when not moving; and even when the rods relax, they are still in tension. Under the worst stretching conditions, however, the rods can sometimes be compressed.

When the upstroke begins at the surface, the pump may still be moving downward; and when the downstroke begins, the pump is still moving upward. This type of pump operation is called *overtravel* because the pull rod stroke length is longer than the polished rod stroke length. Under different stretching conditions, the pull rod stroke length can be shorter than the polished rod stroke length. This latter condition is called *undertravel*.

There are a number of methods which can be used to analyze and predict the behavior of a sucker rod pumping system. These methods account for the complex mathematical relationships needed to describe the reactions of the rod string. Some of these methods only predict the pumping system's operation and are used to select the proper equipment type and size, while other methods are used to analyze the operation of an existing system. These design and analysis methods can be used to predict the pumping rate and the surface and bottom hole pressure of a well.

Analysis and Data Gathering

One of the best ways to determine the operation of a subsurface pump is through the use of a dynamometer (Figure 3-48). This instrument is inserted

86 Introduction to Petroleum Production

Figure 3-48. A dynamometer records polished rod load versus polished rod position.

into the bridle and carrier bar arrangement as shown in Figure 3–49. A dynamometer makes a recording of polished rod load versus polished rod position. A dynagraph (Figure 3–50) is the chart recorded by a dynamometer and may be obtained and analyzed to determine well operation.

Other data that should be gathered from a well are latest pumping rate, surface stroke length, stroke speed, tubing pressure, and casing pressure. These data are used to determine if the pumping system is performing acceptably. A sudden change in operating parameters can indicate a failure in subsurface equipment.

Another measurement that can yield valuable information is the sonic detection of annular fluid level. The level at which liquid stands in the tubing-casing annulus is an indication of bottom hole pressure, because the hydrostatic head of this liquid must equal the bottom hole pressure. Annular fluid level is measured with an acoustic device which determines the time required for an explosive sound to echo from the annular liquid. This time is proportional to the distance from the surface to the liquid.

Artificial Lift 87

Figure 3-49. A dynamometer is inserted in the space provided by spools and plates between the polished rod clamp and the bridle carrier bar.

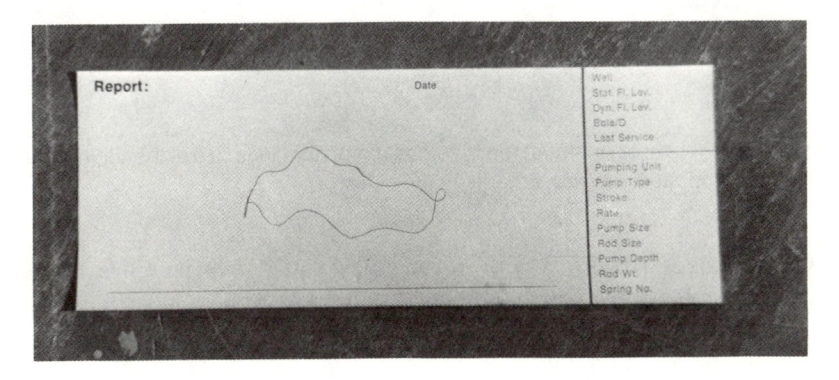

Figure 3-50. The dynamometer card recorded on a dynamometer shows the mechanical operating characteristics of a pumping unit.

Wellhead Arrangement

The equipment used at the surface of a well being pumped with a sucker rod lift system must include provisions for sealing pressure inside the tubing while allowing the polished rod to move freely. The stuffing box accomplishes this objective. Liquid flows out of the tubing through the pumping tee shown in Figure 3–51. This is one arrangement for removing liquid and gas from a pumping well. Another method is shown in Figure 3–52, where gas from the annulus is used to push liquid out of the pumping tee.

Figure 3-51. A simple pumping tee provides a path for fluids to exit the tubing while allowing unimpeded rod movement.

Check valves (Figure 3–53) are one-way valves which utilize a metal wafer that swings up when fluid flows in one direction but blocks flow in the opposite direction. These valves are used on wellheads to assure that flow moves in one direction only.

Chemical treatments for scale, corrosion, or paraffin are often performed at the wellhead. Figure 3–54 shows a wellhead arrangement for using the sucker rod pumping system to pump liquid into a well. When chemical is to be placed in the wellbore, the valves are manipulated, and liquids are circulated down the annulus after being lifted up the tubing. Inhibitors for scale, corrosion, and paraffin or solvents can be pumped into a well in this way. When all the chemical has been pumped, the valves are returned to the normal position, and the well resumes pumping.

Abnormal Operation of Sucker Rod Lift Systems

Sucker rod pumping systems are in normal condition when the pumping unit runs continuously, the rod stretch and relaxation causes no overtravel,

Artificial Lift 89

Figure 3-52. Gas from the annulus passes through a pumping tee and assists in pushing fluid through the flowline.

Figure 3-53. A check valve uses a wafer to allow fluid to flow in one direction but prevents fluid flow in the opposite direction.

90 Introduction to Petroleum Production

Figure 3-54. Tubing pressure can be employed to pump chemicals back into the well by using automatic circulation systems.

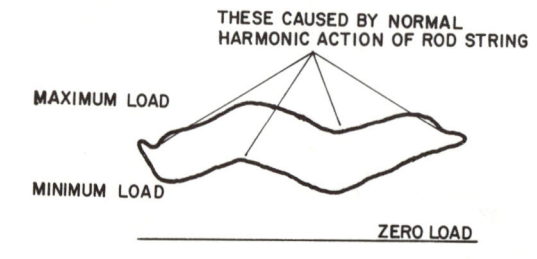

Figure 3-55. A dynagraph is a mechanically produced graph of polished rod load versus polished rod position.

and the pump fills with liquid completely on each stroke. When a dynagraph is made for this operation, it appears as in Figure 3–55. The liquid production rate from such a well should be near the maximum capacity of the pump.

Fluid Pound. When the pumping rate is greater than the rate at which fluid enters the wellbore, the pump can become starved for liquid when only part of the barrel's volume is filled during the upstroke (Figure 3–56). This can also happen when liquid flow into the pump is restricted. The space above the liquid in the barrel is occupied by water vapor or gas that is liberated

Artificial Lift   91

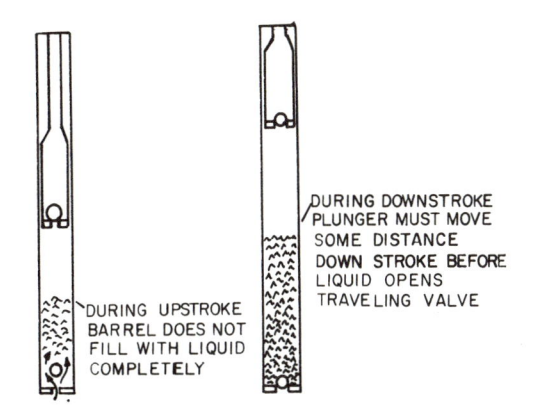

Figure 3-56. Fluid pound occurs when the barrel only partially fills during the upstroke.

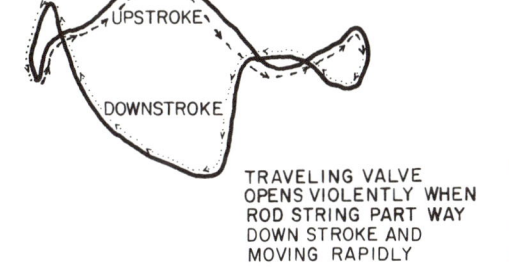

Figure 3-57. Fluid pound causes severe load changes which can damage equipment in the pumping system.

from the oil. Physical law says that a space cannot be totally empty. On the downstroke, the gas is compressed, but the pressure inside the barrel does not open the traveling valve until the valve strikes liquid. By this time, the rods, which had stopped between the up and down strokes, have accelerated. When the traveling valve opens, the weight on the rod string can suddenly drop by thousands of pounds in a fraction of a second. This condition is called fluid pound, and the dynagraph for fluid pound is shown in Figure 3–57. Fluid pound should be avoided because it causes extreme stresses which can result in premature failure of subsurface equipment. Fluid pound can be corrected by slowing the pumping unit, shortening the stroke length, or any other method of bringing pump capacity closer to the flow capacity into the pump.

Gas Interference. In the case of fluid pound gas is in the barrel because the space has to be occupied by something; the gas that boils from the liquid is

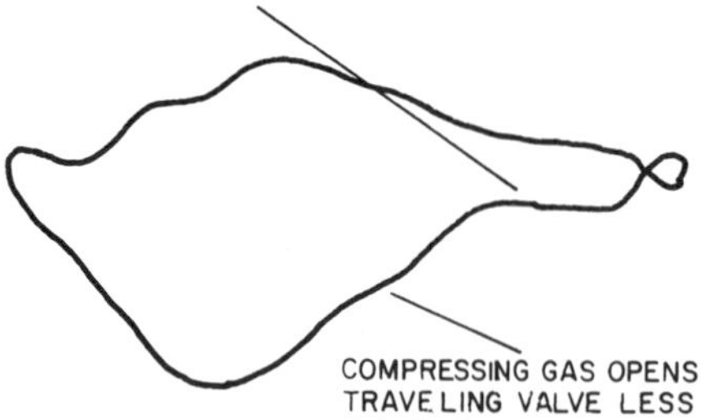

92 Introduction to Petroleum Production

ROD STRING MOVES PART WAY DOWN STROKE BEFORE COMPRESSING GAS OPENS TRAVELING VALVE

COMPRESSING GAS OPENS TRAVELING VALVE LESS VIOLENTLY THAN DOES LIQUID

Figure 3-58. Gas interference reduces lift efficiency but does not necessarily cause equipment damage.

actually the result of low pressure. When gas instead of liquid enters the pump, it does so at much higher pressure. When the downstroke begins, the gas is compressed; but this time the pressure of the gas in the barrel can reach the pressure needed to open the traveling valve before the barrel reaches liquid. The traveling valve opens slowly without the drastic load change experienced in fluid pound as shown in Figure 3–58. This situation, called gas interference, does not normally cause premature equipment failure, but it does indicate poor pump efficiency. Gas interference can be corrected by lowering the pump or by using the bottom hole separator or gas anchor.

Gas Lock. When even more gas enters the pump, a phenomenon called gas lock can occur in a subsurface pump. On the downstroke, pressure inside a barrel completely filled with gas may never reach the pressure needed to open the traveling valve, and whatever was in the pump barrel never leaves it. However, on the upstroke, the pressure inside the barrel never decreases enough for the standing valve to open and allow liquid to enter the pump. Thus, for stroke after stroke, no fluid enters or leaves the pump, and the pump is gas locked. Although this condition poses no real problem in terms of equipment failure, the pump is not functioning at all, and the pumping system is useless.

Bottom hole pressure is determined by flow rate into the wellbore. As flow rate decreases or stops, the bottom hold pressure increases. This increased pressure will support a high column of liquid in the annulus because the hydrostatic head equals the bottom hole pressure, and a decrease in pumping rate is accompanied with an increase in annular fluid level. In many cases of gas lock the bottom hole pressure may increase to the point that it exceeds the pressure in the barrel when the plunger reaches its apex, and liquid can

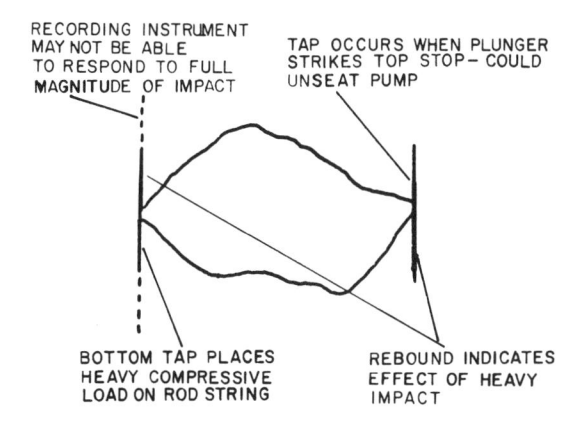

Figure 3-59. Tapping can cause severe damage to the rod string and subsurface pump when they are subjected to excessive stresses.

enter through the standing valve. After a few strokes, enough liquid enters the pump that the gas lock is broken, and the pump functions normally.

Tapping. Subsurface pumps have mechanical stops at the top and bottom that prevent movement past them. These stops are to be used only when a pump is installed or removed from a well. During normal operation, the plunger should never touch either stop. Solids can sometimes enter a pump and interfere with valve operation, and the pumps can gas lock occasionally. It is and has been a prevalent practice for many years for production operators to install a pump such that it intentionally taps the bottom and/or top stops. Tapping the stops at the top and bottom of a stroke jars the pump and causes the balls to jump off their seats in both valves.

It is a fairly well-established fact that such a practice is successful in preventing gas lock. Also, this tapping operation can help in washing sand or other solids out of the valves. However, in terms of equipment operation this is by far the worst possible way to operate a subsurface pump. Figure 3-59 is a dynagraph of a well which is tapping both top and bottom. The load on the rod string is 2-10 times the normal maximum rod load and may exceed the maximum yield of the rods, causing early rod failure. The tremendous jar caused by tapping can damage all working parts of a subsurface pump, particularly the balls and seats in both valves. The benefits gained with tapping are far outweighed by the mechancial damage caused by this practice.

Optimizing Sucker Rod Lift Systems

Selecting the correct equipment for use in a well being pumped with a sucker rod system is a difficult task that requires a great deal of time and consideration and (contrary to common opinion) differs for every well—even when dealing with neighboring wells. Many operators claim that the only way to make a new well "start pumping" is to make it tap so that gas and sediment do not interfere with pump operations. This is not true! By properly selecting and operating the pumping system equipment, these problems can be overcome without intentionally damaging the entire system.

The first step in establishing a proper pumping system is selecting the optimum equipment sizes and operating parameters such as pumping unit size, rotation speed, and stroke length; rod size and configuration; tubing size and anchoring method; and pump type and size. There are several excellent techniques for designing a sucker rod lift system, and one of the best is the API recommended practice RP-11L, an interactive selection technique that has been proved to yield optimum results on many wells.

Most design techniques establish a pumping system ideally matched to a well's flow capacity when the system is operating under stable and continuous conditions. These techniques also assume a pump operating at nearly 100% efficiency. Although most pumps begin operating at this efficiency, normal operating wear, particularly in the valves, reduces this efficiency. Many operators prefer over-designing a pumping system by 20–50% so that the pump can wear significantly and still recover all available fluid without replacement. Unfortunately, until the pump wears sufficiently, fluid pound occurs almost continuously.

There are several solutions to designing for future pump wear. One solution is to use a larger pumping unit than would be normally used. The pumping unit would be initially operated with a short stroke length and at low speed. As the pump wears, the stroke length and speed can be increased (a simple mechanical adjustment at the surface) to compensate for pump wear.

Another more cost-efficient method is to over-design the system and use a timing system or electronic control device to operate the pumping unit intermittently. When the pumping system is operating, its capacity is more than that of the reservoir, but the system is stopped before fluid pound, gas interference, or gas lock can develop. While the system is stopped, bottom hole pressure and annular fluid level increase so that when the pumping unit starts, fluid pound, gas interference, or gas lock will not occur.

For a subsurface pump to operate properly, it must be submerged in liquid. To achieve the best operation of a sucker rod pumping system, it should be configured as shown in Figure 3–60. If the pump is placed below the productive intervals, gas is allowed to flow upward in the annulus, while liquid falls to the bottom of the well near the pump. This configuration can virtually eliminate gas lock, gas interference, and fluid pound. Tapping is not needed in this sytem as long as pump submergence is adequate.

The wellbore of a producing well is said to be the best possible gas separator, but sometimes it is not possible to place the pump below the productive

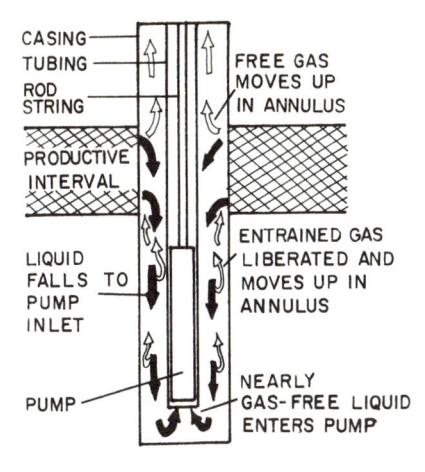

Figure 3-60. A subsurface pump should be installed below producing intervals to obtain the best gas separation in the annulus.

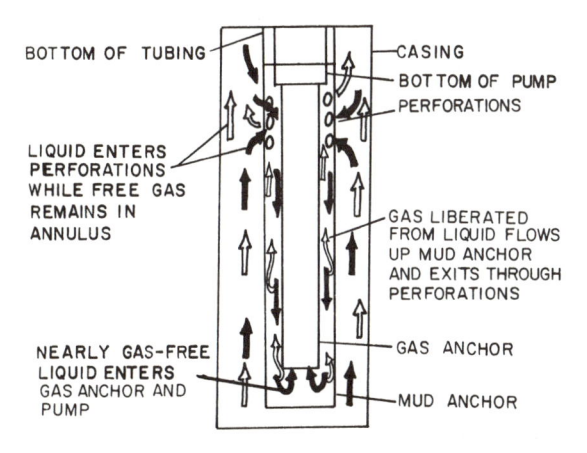

Figure 3-61. When a pump cannot be installed below the producing intervals, a mud anchor/gas anchor should be used to prevent gas from entering the pump.

intervals, or liquid enters so rapidly that it carries gas downward to the pump inlet. In such cases a bottom hole configuration allowing gas separation (Figure 3–61) is used. The separator or gas anchor can be used effectively to remove gas from liquid if it is designed to allow unrestricted liquid flow. Flow restrictions can cause fluid pound.

Solids in a wellbore can accumulate and plug separators, gas anchors, and the internal parts of the pump. Sometimes, such as immediately after a fracture stimulation, pumping solids cannot be avoided. Then three-tube pumps should be used to remove sand from the well but should be replaced with high-efficiency pumps as soon as possible. Some operators keep used pumps—in which efficiency is already low—to install in a well long enough to clean out the sand. Again, the clean-out pump is replaced as soon as possible.

Solids such as formation sand, scale, or paraffin may present permanent obstacles to sucker rod lift systems. However, these problems can be overcome without resorting to improper operating practices. Reservoir solids can be stopped by completion practices such as using sand-gravel packs or sand control chemicals. Scale and paraffin deposition can be controlled or totally inhibited by periodic chemical treatments and by selecting proper reservoir withdrawal rates.

Virtually all sucker rod lift problems can be overcome without resorting to techniques such as tapping or intentional fluid pound which can cause equipment damage. The equipment used for sucker rod pumping will eventually wear out, but if properly installed and operated, this equipment can operate satisfactorily for more than 10 years. In the past it was not critical that wells be properly operated because the few hundred dollars to repair a broken rod or damaged pump was insignificant compared to the time and cost required to operate a system properly. This is no longer the case because even a simple broken rod replacement can cost thousands of dollars.

Most sucker rod lift system equipment is rated for 10–15 million reversals (strokes); however, when operated under the best conditions, several times this life can be expected. In most cases the best possible operation can be obtained by using the longest possible stroke length and the slowest possible speed to attain a given production rate. This places the least stress on the entire system and prolongs the life of all operating parts.

Electric Submersible Pumping

Another form of artificial lift uses a self-contained electric motor, seal assembly, and centrifugal pump attached to the tubing string as shown in Figure 3–62. Electricity is supplied to the motor by a shielded, armored cable. Submersible pumps are particularly effective for very high pump rates of several thousand barrels per day, but they are applicable for almost any pumping application.

Subsurface Pump/Motor Assembly

The subsurface unit of a submersible pump is shown in Figure 3–63. It consists of the pump, a gas separator (if used), the seal section, and the electric motor. The motor is installed below the pump because fluid must flow upward past the motor to carry away the motor heat.

Subsurface Motor. The motor used to drive the pump is a three-phase, squirrel-cage induction motor designed and constructed for this application.

Artificial Lift 97

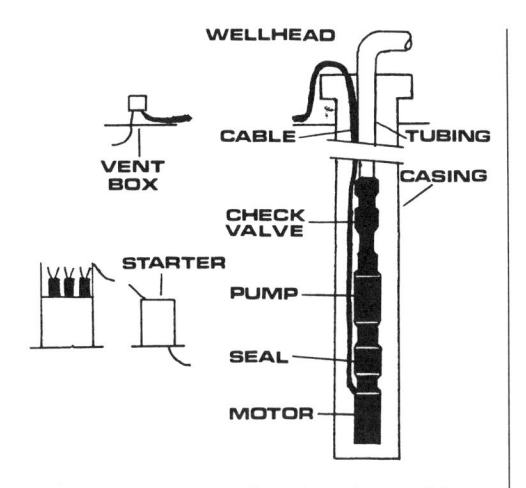

Figure 3-62. An electric submersible pumping system consists of surface electrical switch equipment and cable and a subsurface electric motor and centrifugal pump to lift liquid to the surface. (Courtesy of Centrilift Hughes.)

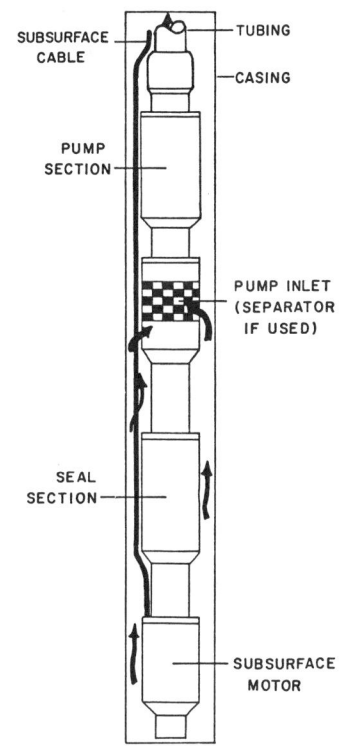

Figure 3-63. The subsurface motor-seal-pump assembly must be mechanically sound to withstand subsurface pressures and temperatures as well as the rigors of installation.

Figure 3–64 shows the motor for a submersible pump. The motor may consist of several sections stacked on each other. Large motors of 50 horsepower or more are usually used, and the only cooling possible is the reservoir liquid moving upward past the motor. The motor is designed to seal the high bottom hole pressure out. To keep the current-carrying conductor size small, the motors are designed for high voltage of several thousand volts from one conductor to another. Special lubrication is required on these motors because they operate at several hundred degrees Fahrenheit—temperature that will thin most lubricating oils. A specially formulated, high-temperature oil is permanently sealed in the motor. Because of the high voltages and the unique lubrication requirements, the motors are completely sealed against fluid entry. The slightest leak can damage the motor very quickly.

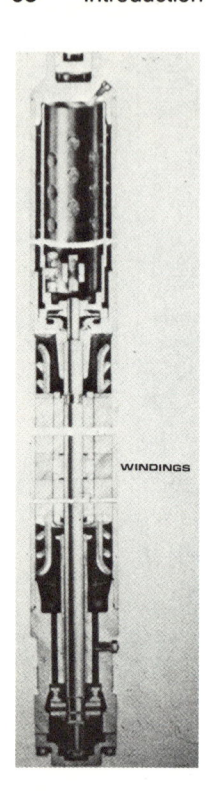

Figure 3-64. The subsurface electric motor is designed to deliver great power from a compact unit. (Courtesy of Centrilift Hughes.)

Seal Section. The seal section of a submersible pump (Figure 3–65) accomplishes two major tasks. First, it provides a means by which the motor shaft can turn the pump but prevent fluid entry. Second, the seal section allows the pressure of oil inside the motor to be the same as that outside. This minimizes the tendency for wellbore fluids to enter the pump.

Pump Section. The pump section of a submersible pump consists of many individual impellers (each like a small centrifugal pump) attached to a single shaft. The operation of a single impeller is shown in Figure 3–66. Each stage of a submersible pump adds a little pressure to the liquid stream. By the time fluid passes from the bottom to the top of the pump, the fluid pressure has been raised above the hydrostatic head of the tubing fluid so that fluid is pushed up and out of the tubing.

A complete pump section consists of many individual pump impellers as shown in Figure 3–67. The bottom impellers are designed for low suction pressure, while the top impellers have high suction and discharge pressures. When gas is expected to be pumped, the lower elements, called *compression stages*, must raise the pressure of gas and liquid rapidly so the gas will dissolve in the oil. The upper stages are designed for flow of compressible liquid.

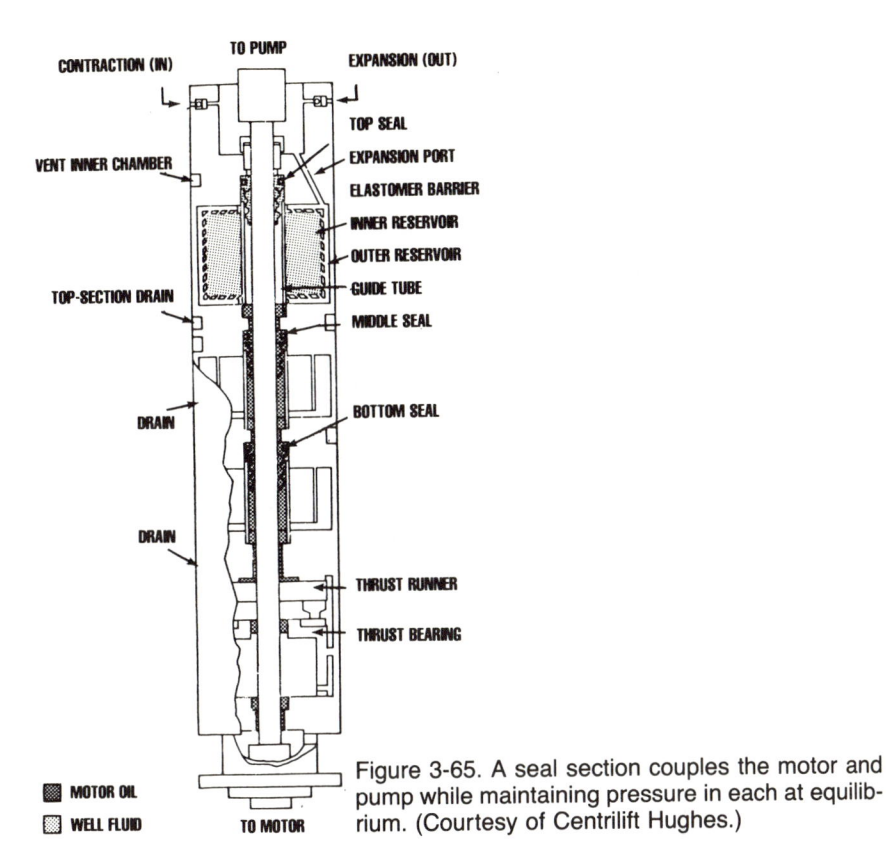

Figure 3-65. A seal section couples the motor and pump while maintaining pressure in each at equilibrium. (Courtesy of Centrilift Hughes.)

The efficiency of any centrifugal pump is much lower than that of a positive displacement pump because some fluid always slips by the impellers. Gas entry into a submersible pump decreases the lift efficiency even more.

The configuration of a submersible pump must be designed for every well individually. The pump design must take into consideration the inlet pressure, the discharge pressure required to lift liquid to the surface, the friction loss in the tubing, the amount of gas dissolved in the oil, the amount of free gas, the pumping rate, and fluid characteristics such as density and viscosity.

Gas Separator. Although submersible pumps can be designed to handle free gas, the efficiency of such designs is normally low, and the pump is expensive. A gas separator can be placed at the pump suction to help remove much of the free gas before it can pass through the pump. A gas separator, pictured in Figure 3–68, functions much like the bottom hole gas separators used with sucker rod pumping systems. Fluid is made to flow through a serpentine path where gas separates from liquid and leaves the separator while liquid passes through the separator to the pump inlet.

100 Introduction to Petroleum Production

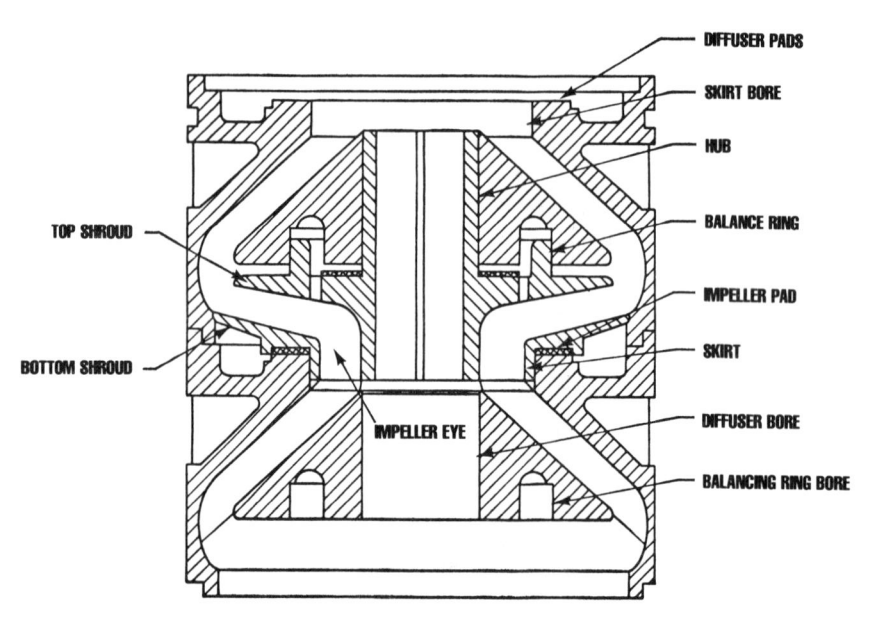

Figure 3-66. A pump impeller lifts liquid by slinging it upward as the shaft rotates. (Courtesy of Centrilift Hughes.)

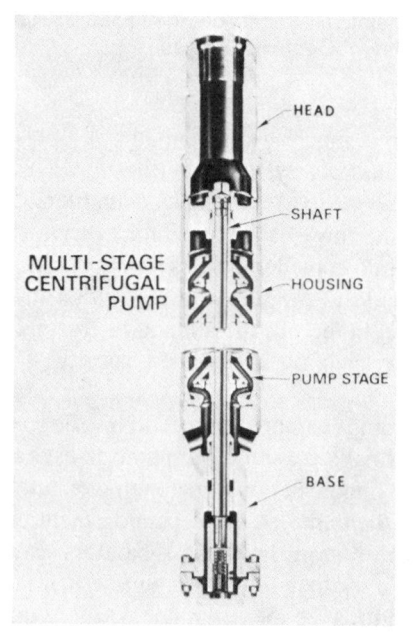

Figure 3-67. A complete pump consists of many impellers, each moving liquid up to the next. (Courtesy of Centrilift Hughes.)

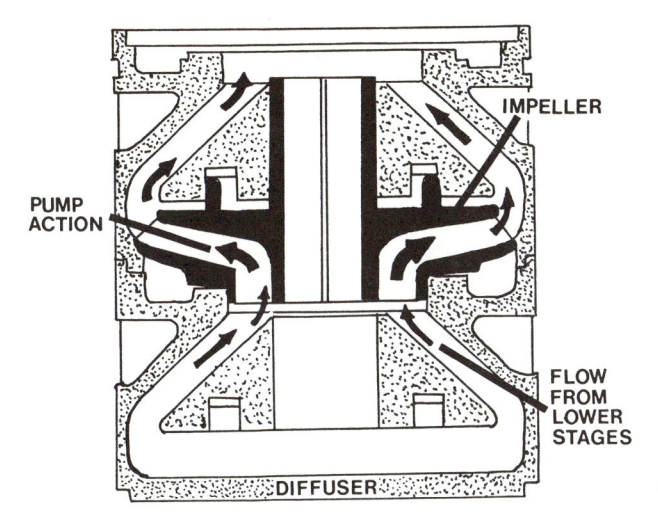

Figure 3-68. The separator allows liquid to flow unimpeded into the pump while preventing gas from entering the pump. (Courtesy of Centrilift Hughes.)

Motor Shroud. Motors can be damaged very quickly if they are not cooled. Sometimes liquid flow past the motor, as shown in Figure 3–63, is not enough to cool the motor adequately. In such cases a motor shroud (Figure 3–69) is used to channel all liquid past the motor so that the liquid can flow very rapidly. This aids in cooling the motor, but the restriction to flow must be considered when designing the pump. Also, the outside diameter of the shroud is so large that it may not fit in the wellbore.

Subsurface Cable

The cable used to supply electric current to the motor (Figure 3–70) must be armored cable with excellent corrosion-resistance and insulating qualities. High-power motors (about 50,000 watts for a 50-horsepower motor) require either high voltage or high current. Large currents require large-diameter copper conductors (0.75 inches in diameter), but physical law allows high power to be transmitted with low current when high voltage is used. Most submersible pump manufacturers use high voltage so that lower currents can be used. The lower current requires smaller conductors of less than 0.25 inches in diameter, and this size is important in keeping cable size small.

The high voltage requires that the cable have excellent insulating properties—even the slightest insulation breakdown can cause the cable to short out. The plastic and rubber normally used for electric wire is not satisfactory because it is attacked and damaged by wellbore fluids. The cable must be covered with steel armor to prevent mechanical damage. A thin metallic tape

102 Introduction to Petroleum Production

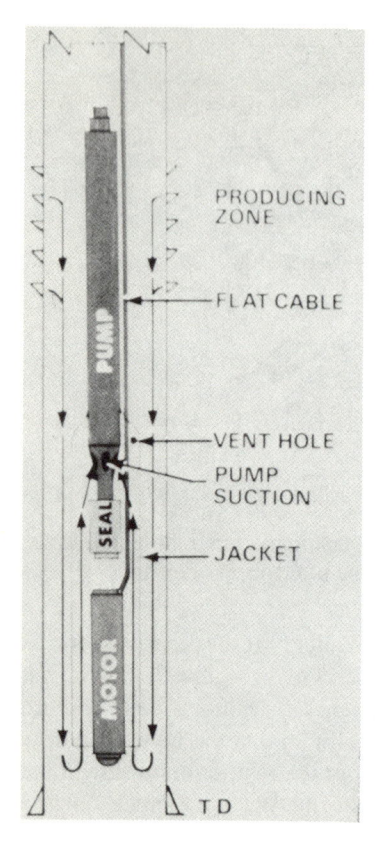

Figure 3-69. A motor shroud or jacket forces liquid to flow by the motor before entering the pump to provide cooling for the motor. (Courtesy of Centrilift Hughes.)

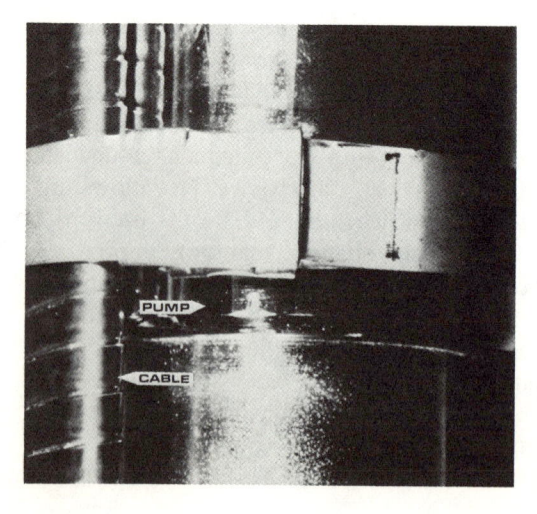

Figure 3-70. Armored electric cable is clamped to the tubing string as well as to the pump to prevent damage. (Courtesy of Centrilift Hughes.)

or braid, imbedded in the insulation, shields the insulation from the electrical stresses associated with high voltage.

For many years the weakest part of a submersible pumping system was the cable. Rubbers and plastics were invaded by wellbore fluids, particularly hydrocarbon gases, and the insulating properties were destroyed. The high pressures in the well aggravated this problem. Recently, vast improvements have been made in compounding the insulation used in subsurface cable. Also, improvements have been made in the cable armor. The improvements in cable construction have eliminated many of the problems associated with cable. In fact, premature failures of virtually all equipment in submersible systems have been reduced to the point that such systems are now considered almost as reliable as sucker rod systems.

Subsurface cable still can and does fail occasionally. When a pump stops operating, a simple test can be made at the surface to determine if a cable has broken down. If it has, the cable is removed from the well, the bad part of the cable is cut out, and the remaining cable is spliced and placed back in service. Splices should be made in such a way that their mechanical and electrical qualities equal or exceed those of the original cable.

Installation of Subsurface Equipment

A submersible pump is attached to the production tubing itself so that the tubing must be removed from a well every time the pump is serviced. This requires a well servicing unit to manipulate the tubing, but it also requires special cable reels and tools to reel cable into a well or remove it from the well. Figure 3–71 is a photograph of the surface equipment required to install a submersible pump.

The pump, seal, motor, shroud, and separator are assembled as a complete unit before installation. One end of the cable is terminated through a special entry bushing to the motor. The bottom hole assembly is then attached to the bottom joint of tubing. As upper tubing joints are attached and lowered into the well, the cable is clamped to each. Great care must be exercised to assure that the cable is not damaged as it is lowered into the well. Also, the subsurface pump equipment (particularly the shroud) can be easily damaged by obstructions in the wellbore. One advantage of a submersible pump is that no other subsurface equipment—including anchors and packers—is required

Surface Equipment

The equipment configuration of a well being pumped with a submersible pump is much the same as that in a flowing well. Figure 3–72 is a typical arrangement. The cable is brought out of the tubing-casing annulus through

104 Introduction to Petroleum Production

Figure 3-71. A cable reel is required to handle cable while an electric submersible pump is installed or removed. (Courtesy of Centrilift Hughes.)

Figure 3-72. A wellhead of a well equipped with an electric submersible pump must be configured to handle the cable and may be adapted for chemical injection or other functions.

a service cable entry bushing, a rubber-packed sealing mechanism that keeps well fluids inside while allowing the cable to come to the outside.

The cable is terminated in a motor starter. Because the motor uses high voltage, the motor starter and electrical switchgear are larger than those used for most electric motors in a production operation. The starters are also equipped with several instruments which analyze the operation and performance of the electric motor.

Many wells using submersible pumps are equipped with chokes or other control valves to regulate the flow of liquid from the tubing. It is sometimes necessary to control surface pressure to assure proper operation of the pump. Because both tubing and casing pressure are important in proper operation of the pump, gauges, recorders, or electronic monitoring devices may be installed to warn if abnormal conditions develop.

Hydraulic Pumping Systems

Another major artificial lift system presently in use is the hydraulic pumping system. Hydraulic pumping systems were used extensively until the mid-1970s, but there are still applications when hydraulic pumping systems may be used effectively. A hydraulic engine/pump assembly is put at the bottom of the production tubing. High-pressure liquid is pumped down the power fluid tubing to operate the hydraulic engine. Figure 3–73 is a schematic representation of a hydraulic pumping system.

Subsurface Engine Pump

Hydraulic Engine. Figure 3–74 is a schematic representation of a hydraulic engine and pump. During the upstroke, the high-pressure fluid is routed below the engine cylinder. The high pressure works against the cross-sectional area of the cylinder to exert considerable force on the connecting rod. When the engine cylinder reaches the top of the stroke, the engine valve switches, routes power fluid above the cylinder, and pushes it down. The connecting rod applies the force generated by the hydraulic engine to the pump cylinders.

Hydraulic Pumps. The pump attached to the engine may be identical in operation to a sucker rod pump. That is, liquid is lifted on the upstroke, and the pump returns to its starting position on the downstroke by gravity or with hydraulic assistance. This pump arrangement is called a *single-acting pump*.

A *double-acting pump* allows fluid lift on both the up and down strokes. On each upstroke, check valves allow fluid entry in one chamber and exhaust in the other. On the other half of a full stroke, the chambers reverse operation.

106 Introduction to Petroleum Production

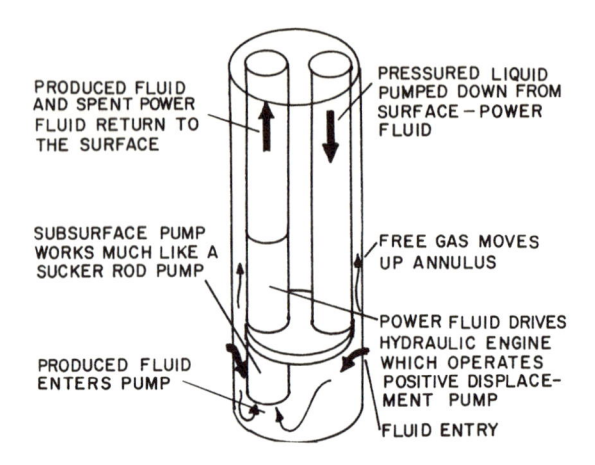

Figure 3-73. A hydraulic pumping system lifts liquids with the energy from high-pressure liquid pumped down from the surface.

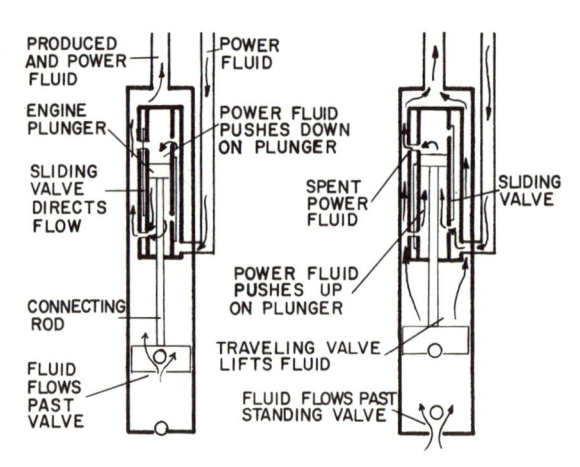

Figure 3-74. A hydraulic engine converts the high pressure from the power fluid to the reciprocating movement needed to operate the pump.

Subsurface Unit. The hydraulic engine operates with short strokes of a few inches. To obtain the high rates needed, the engine must operate at high speeds of 50–150 strokes per minute. The engine and pump are made of finely machined steel parts that have very close tolerances. Because of the high speed involved, the engine and pump parts are subject to failure. Repair of the precision parts of the pump can be very expensive, so it is important that the engine and pump be operated properly to avoid premature equipment failure.

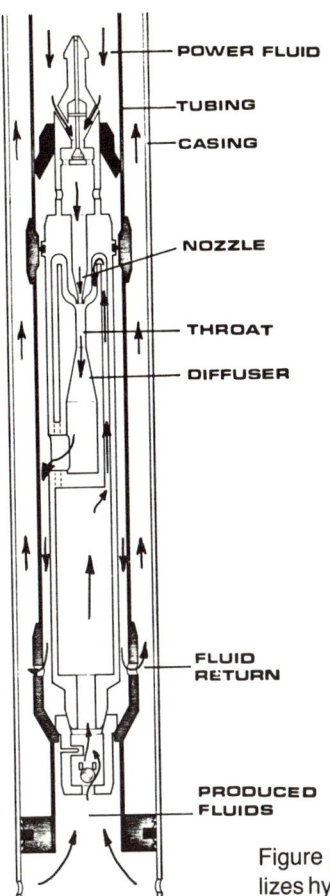

Figure 3-75. The jet pump is a subsurface pump that utilizes hydraulic principles instead of moving parts to lift fluid. (Courtesy of Kobe, Inc.)

Power Fluid. The high-pressure liquid used to drive the hydraulic engine is called power fluid. Either oil or water may be used, but most users prefer slightly compressible oil rather than water because the severity of water hammer is reduced somewhat, and oil is usually less corrosive than water. Because the engine uses close-tolerance machined parts, it is absolutely essential that the power fluid be clean and completely free of entrained solids.

Hydraulic Jet Pump

A recently developed subsurface pump is the jet pump (a product of KOBE, Inc.) which is shown in Figure 3-75. In a jet pump the power fluid is forced to flow through a small nozzle into a venturi. Here, the high-pressure power fluid stream is converted to a high-velocity, low-pressure stream. Produced fluid flows into this low-pressure area and is accelerated by the high-speed power fluid. In the diffuser the high-velocity, low-pressure stream is converted back to a low-speed, high-pressure stream. The pressure in the diffuser has enough hydrostatic head to push both liquids to the surface.

Control of Hydraulic Pump Rates

The only method of controlling the rate at which a hydraulic pump operates is to adjust the stroke speed. Adjustments are made by regulating the rate at which power fluid enters tubing at the surface. Jet pump rates are also controlled by modifying surface power fluid rates.

The wellhead of a hydraulic pumping system is equipped with pressure gauges. The gauge on the production tubing string oscillates as the pump operates. The number of pump strokes per minute can be counted from this fluctuation. The pump rate is adjusted by opening or closing the control valve and then checking the gauge to see that the desired rate has been attained.

Power Fluid Conditioning

The first task of a power fluid conditioning facility is separating gas from liquid. Power fluid must be almost incompressible and certainly cannot contain free gas which would gas lock the engines. Thus, all free gas and as much solution gas as possible must be removed from the liquid used for power fluid.

In addition to separating gas from liquid, it is also necessary to separate oil from water before liquid can be used as power fluid. Water or oil can be used for power fluid, but a mixture of oil and water is not usually acceptable for power fluid.

Power fluid is raised to the necessary high pressure by high-pressure, high-rate, positive displacement pumps. The pumps used for this application are similar to those used for many other liquids and are discussed in Chapter 6.

Centralized Power Fluid Facilities. When hydraulic pumping systems were introduced, it was economically attractive to use a single facility to prepare power fluid for use in the wells. Such a facility is shown diagramatically in Figure 3-76. More recently, centralized facilities have been replaced by individual power fluid systems to reduce installation and operation costs.

Solo Power Fluid Conditioning Units. On many leases there are not enough wells using hydraulic pumps to justify the high cost of central power fluid conditioning facilities. In the last few years low-cost, self-contained

Artificial Lift 109

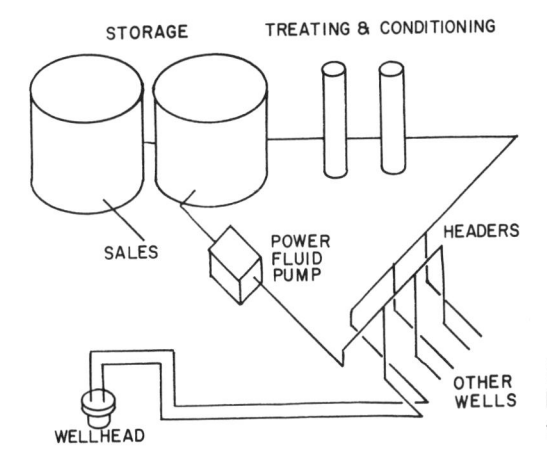

Figure 3-76. Power fluid can be provided to several wells from a central facility.

Figure 3-77. A solo hydraulic pumping system uses a skid-mounted unit containing separation, cleaning, storage, pumping, and control equipment required to operate a subsurface hydraulic pump.

power fluid conditioning units have been developed to offset the high cost of central facilities. These are skid-mounted systems which contain separation, cleaning, storage, and pumping equipment. One such unit is shown in Figure 3-77.

Maintenance and Operation

The components of a hydraulic pumping system are close-tolerance machined steel parts which are designed for high-speed operation. These parts are often made of stainless steel or other hardy materials to inhibit corrosion.

The components of most parts of the hydraulic pumping system, particularly those of the subsurface engine/pump assembly, are expensive to manufacture, and replacement costs can be very high. Also, highly skilled service technicians are required to repair this equipment. The use of water as power fluid is detrimental to all subsurface and surface equipment. The maintenance costs of hydraulic systems are the highest of any of the artificial lift systems. Many hydraulic pumping systems are replaced with sucker rod and electric submersible pumping systems, both of which have significantly lower operating costs.

Artificial Lift System Analysis

In every artificial lift method there is some parameter that can be measured that reflects the amount of energy expended in the lift system. In gas lift and plunger lift systems the amount and pressure of injected lift gas is an indication of the amount of energy being used for lift. In sucker rod pumping systems the dynamometer shows the amount of energy being used and allows analysis of subsurface equipment operation. Much the same information can be gained by measuring the motor current when an electric motor is used as the prime mover of a sucker rod pumping system. The motor current for an electric submersible system is also used for such input energy measurements. The flow of power fluid in a hydraulic pumping system yields the same information.

When the rate of fluid production is measured and compared to the input energy stream, the lift efficiency of the system can be determined. The best lift system design is the one which makes the most efficient use of lift energy.

Effect of Natural Flow on Artificial Lift

In some wells being pumped using one of the artificial lift methods the reservoir may still have enough pressure to flow without assistance. When gas lift is being used, natural flow has little effect and is usually desirable. When positive displacement or centrifugal pumps are being used in a well, the effect of natural flow depends on the depth at which the pump is set. Since gas is almost always associated with natural flow, the gas can enter the pump if it is set above the productive interval. Gas entry into a centrifugal pump has the effect of reducing efficiency but causes little damage to the pump. On the other hand, gas in a positive displacement pump can cause severe problems. If the pump is set below the interval from which flow occurs, the reservoir fluids will probably flow up the annulus without much effect.

Artificial Lift

Natural flow can occur in a well sporatically. An oscillatory pressure action may take place in a reservoir, and a well may be produced solely by the artificial lift system for a time and later by natural flow in periodic cycles. In other wells the artificial lift system may be used only to augment natural flow.

Casing Vacuum Pumps

In some wells even the slightest annular pressure can restrict flow of reservoir fluids. Even intentionally over-designed pumping systems cannot reduce bottom hole pressure enough to maximize reservoir fluid flow.

A vacuum pump's inlet pressure is intended to be lower than atmospheric pressure, whereas most pumps depend on inlet pressure at or above atmospheric. When a vacuum pump is used on the casing of a well, its only purpose is to reduce the pressure in the annulus. Casing vacuum pumps are used when gas flows into the annulus so rapidly that the surface pressure becomes too high. Because atmospheric pressure is only 14 psi above that of absolute vacuum, vacuum pumps can reduce pressure only slightly.

Comparison of Artificial Lift Systems

When it becomes necessary to use artificial lift systems, a decision must be made as to which type to use. Each system requires a capital expenditure which may be enormous. Also, each system requires certain periodic operational adjustments as well as periodic maintenance. The costs of operation and maintenance for many years can often far outweigh even very large initial capital outlays. Thus, all facets of artificial lift systems must be considered before selecting one.

Gas Lift

Gas lift is often the first artificial lift system used on larger leases. One major reason is that if centralized compression facilities are used, the initial investment and operating and maintenance costs are lower than for any other lift system. The equipment used is simple and virtually maintenance-free. This is the major advantage of gas lift systems.

Because liquid slips downward past the lift gas, gas lift is inefficient. Such a lift system is limited to shallow wells of 2000 feet or less, or to wells which have enough reservoir pressure that the gas lift assists natural flow more than causing it. Deep wells or wells that have low reservoir pressure are not usually good candidates for gas lift.

Sucker Rod Lift Systems

Sucker rod systems are by far the most predominant of the artificial lift systems. They are used on more than 85% of the wells in the continental United States. Sucker rod lift is the oldest form of subsurface pumping—predating the Christian Era by many centuries.

When sucker rod systems are properly designed, installed, operated, and maintained, they can be the most dependable of lift systems. They can be used on the vast majority of wells. Because of the number of sucker rod systems in use, more research, design, and testing effort is devoted to this field than to any other.

There are certain wells which are not as well suited to sucker rod lift systems as to other methods. Ten thousand feet is generally accepted as the maximum depth at which sucker rod systems may be installed and operated to economic advantage. Below this depth or when extremely high rates are expected, other lift systems are usually better choices, because sucker rod systems with this capacity are extremely expensive.

Electric Submersible Pumps

Primarily because of maintenance costs, submersible pumping systems have not been used extensively until recently. The cable has traditionally been the weakest part of these systems, and frequent cable failures cause high maintenance costs. After a great deal of research, cables have been developed which can stay in the hostile environment of wellbores for years.

Submersible pumps are best suited for very high rates (thousands of barrels per day), and for deep wells of more than 10,000 feet with high rates and large volumes of gas. As some of the enhanced recovery methods operate for several years, submersible pumps are likely to become more predominant because they can be designed to handle the anticipated high water and gas flow rates.

Submersible pumps are physically large (5–10 inches in diameter), and there are many wells in which they cannot be used. This is particularly true when a motor shroud is required, since the shroud increases the size of the subsurface equipment.

Hydraulic Pumping Systems

Hydraulic systems have been used for subsurface pumping since the mid-1940s. They have, however, earned the reputation for being the most expensive systems in terms of maintenance costs because the pumps traditionally fail frequently. One of the main reasons for frequent failure of subsurface equipment is improper operation. Hydraulic pumps can be virtually destroyed in a matter of minutes by fluid pound and gas lock. Another common source of operational problems is contaminated power fluid which can quickly damage surface as well as subsurface equipment.

Hydraulic pumps can be used for any liquid pumping application and are acceptable in any well where adequate pump submergence and gas separation can be attained. These pumps are well suited for deep wells because they can be operated at virtually any pressure and temperature. They are also well suited for high-rate pumping in small wellbores because so many different variations in tubing configurations are available.

Fluid Transportation

All artificial lift systems have a common goal: to bring fluids from the bottom of a well to the surface as economically as possible. A great deal of the operating effort and expense of petroleum production is devoted to this important phase, but once fluids have been brought to the surface, something must be done with them. The next phase in processing petroleum is transporting the fluids to production facilities where they can be prepared for sale or consumption. The fluids are usually pumped from a well into a gathering system which collects and distributes the fluids to process facilities. The gathering system and the equipment required is discussed in Chapter 4.

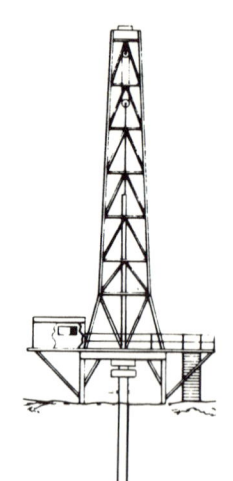

Chapter 4
Surface Gathering Systems

Hydrocarbons must be separated from each other and from water before they may be processed into usable petroleum products. The equipment used for field processing is expensive and is often installed so that several wells are served by a single process facility. The fluids produced from one or more wells are collected in a gathering system and transported to the separation facilities. The gathering system may consist of a single flowline from a well to its separation equipment or many flowlines, headers, and process facilities.

Many states require that wells be drilled according to specific geographical spacing. Depending on ownership of mineral rights and regulations, wells are drilled within areas called leases. The petroleum produced from different leases must be kept separate. Lease restrictions and economics determine the arrangement of gathering systems and the equipment used.

Types of Gathering Systems

One type of gathering system (shown in Figure 4–1) is a radial gathering system. The flowlines in this system converge at a central point where facilities are located. Flowlines are usually terminated at a header, a pipe large enough to handle the flow of all flowlines.

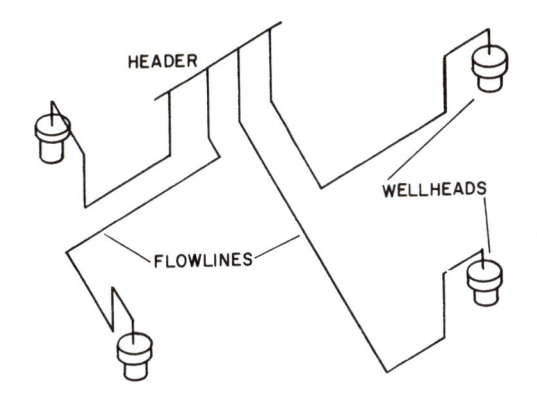

Figure 4-1. In a radial gathering system all flowlines bring fluid to a central header.

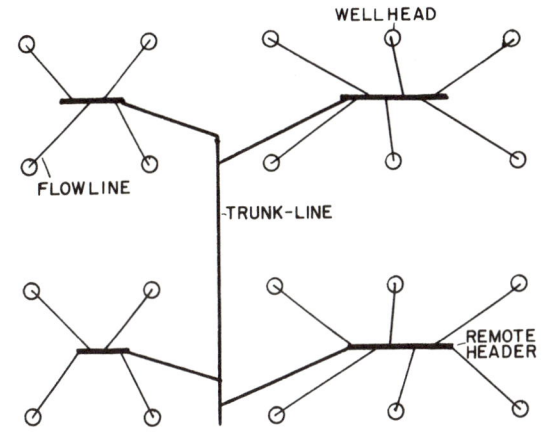

Figure 4-2. An axial gathering system uses remote headers to gather fluid and send it to treating facilities through trunk lines.

Another gathering system is an axial or trunk-line gathering system (Figure 4-2). This gathering system is usually used on larger leases, or where it is not practical to build the process facilities at a central point. The remote headers are simply smaller versions of those used in radial systems.

Leases are equipped to process fluids through equipment large enough to handle all wells simultaneously. To measure the production of individual wells simultaneously, a very complex process and metering facility is required. However, unless some method is provided, all fluid is measured at once; no information on individual wells can be gained.

Regulations and good operating practices require that oil, gas, and water production rates be measured for individual wells. Most leases are equipped so that all but one well on a lease are routed through the large process equipment, and the production from this equipment represents the entire lease.

116 Introduction to Petroleum Production

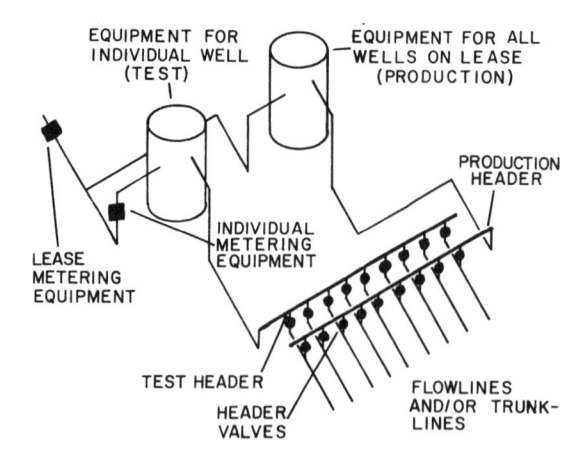

Figure 4-3. By opening and closing selected header valves, an individual well's fluid flow can be measured in a vessel equipped with metering equipment.

The fluids from one well are routed through other equipment so that the flow rates for that well may be measured singly. Such a scheme is shown in Figure 4–3.

To measure the fluid from a well separately, a method must be provided for routing fluids to the production facility (the equipment used for all wells at once) or the test facility (the system for a single well). This selection is made in a well test header, which is diagrammed in Figure 4–4. To test a single well, the test valve is opened and the production valve closed. All other well's production valves are open and test valves closed.

Figure 4–4 is the well test header for a radial gathering system. Figure 4–5 shows a typical remote well test header which is used with an axial system. When test and production facilities are used with a trunk-line gathering system, both test and production trunk lines are required.

The type of gathering system used on an individual lease is determined on the basis of economics. A radial system requires many feet of comparatively small pipe for flowlines. The cost of this pipe may be small in comparison to the cost of the larger pipe required for trunk lines, and the total cost of many long flowlines may be less than that of the same number of short flowlines and one or two long, large trunk lines. The latter is particularly true of the thick pipe used for high-pressure gas production leases.

Another consideration in determining the type of gathering system is the use to which the surface land is put. If the land is rugged pasture land, the flowlines may be laid on the surface and left there permanently. If the land is under cultivation or a populated area, the pipe must be buried 3–5 feet deep

Surface Gathering Systems 117

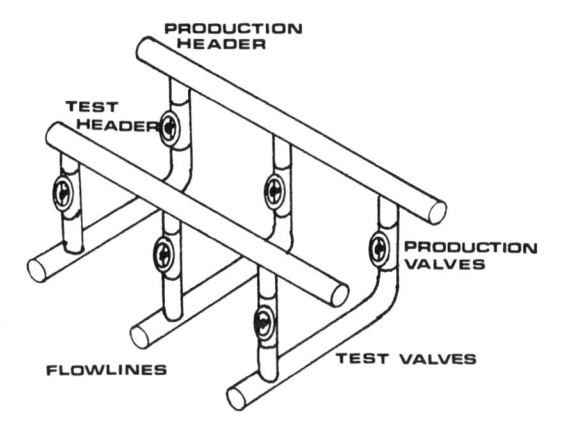

Figure 4-4. A well test header has valves for each well to provide for routing produced fluids to production or test vessels.

Figure 4-5. A remote header is used to select a well for testing, although the header may be far from its associated test and production vessels.

to assure it is below the maximum operating depth of earthworking equipment (like farm machinery). In these areas the flowlines may be required to be buried beneath or adjacent to roads or streets. The length, installation cost, and maintenance costs of the gathering system are influenced by land use, and these factors determine the gathering system type.

Fluid Flow Behavior

The types of flow discussed in Chapter 1 affect the friction in pipe. Of the flow types discussed, only slugging flow normally causes damage to the gathering system. Water hammer occurs when a mass of liquid strikes a sharp turn in a pipe. In slugging flow liquid flows in short bursts, and the changing flow causes water hammer. The gathering system should have gradual rather than sharp turns. Where sharp turns are necessary (in headers for example) the pipe should be adequately supported and braced to prevent damage. Water hammer does not punch holes in pipe from the inside, but the movement it causes can break even the strongest steel after some time of continued fatigue.

Turbulence also increases the friction in pipe. When liquid flow is laminar, the friction loss is significantly less than when flow is turbulent. Although it is not always possible, liquid flow should be laminar throughout the gathering system.

As liquid moves in pipe, the force of friction is exerted against the liquid. The friction opposing flow appears as difference in pressure from one end of the pipe to the other. This pressure difference is commonly called *friction loss* and is influenced by the type of flow, the fluid viscosities, pipe composition, and laminar or turbulent flow. Increasing pipe diameter lessens friction, and it is good practice to implement a gathering system using the largest pipe economically feasible.

When paraffin or scales accumulate in a gathering system, they reduce the pipe diameter and increase friction. The restriction may be so severe that the pipe is plugged. Accumulations of this sort can be prevented by use of continuous chemical solvent treatments, and the cost of the treatments is sometimes less than the expense of cleaning the gathering system. However, when scale or paraffin problems are not bad enough to warrant continuous chemical treatment, the accumulations are allowed to form. The solid material is then removed periodically with strong chemical solvents and/or heat. The criterion for using continuous or periodic treatments is largely economic.

Flowlines

Fluid is transported from wells to treating equipment by pipelines called flowlines. A flowline is connected to a well as shown in Figure 4–6. It is routed underground or on the earth's surface to the equipment needed for field processing. The flowline is terminated in a vessel when only one well is served by the facility or in a header when several wells are involved. The internal flowline pressure is determined by wellhead and process facility pressures, and this pressure determines the materials and construction techniques used to install the flowline.

Figure 4-6. A flowline is connected to a pumping tee and provides a path for fluids to flow to surface production equipment.

Low-Pressure Flowlines

Many wells, particularly those using artificial lift systems, maintain flowline pressure at less than 125 psi. There is no specific pressure below which a flowline is said to be a low-pressure pipeline; however, flowlines with maximum operating pressure at or less than 125 psi are generally accepted as low-pressure lines and are constructed accordingly.

Steel Line Pipe. This is the most common tubular material for flowlines. Line pipe is similar to tubing, except that it seldom has upsets. Line pipe may be obtained with or without threads. The pipe must be cut and threaded so that the flowline matches the dimensions of the wellhead or header.

Anytime the flowline changes directions, or diameter, or is connected to equipment, pipe fittings must be used. Figure 4–7 shows some of the pipe fittings available for use with steel pipe.

The size of line pipe is its outside diameter. Pipe sizes range from one-quarter inch to about 10 inches in diameter. Larger line pipe is available, but this pipe is usually not threaded.

Line pipe is usually specified by the American National Standards Institute (ANSI) schedule numbers. The schedule number indicates the pipe's minimum yield rating in 1000-psi increments. For example, Schedule 40 pipe has a minimum yield rating of 40,000 psi. The most common pipes are Schedule 40, 80, and 100.

120 Introduction to Petroleum Production

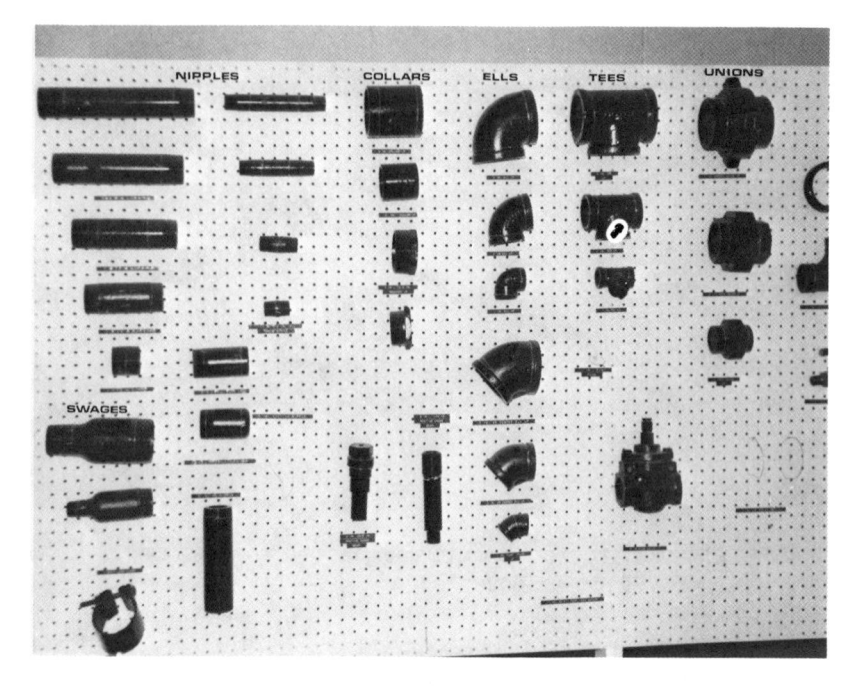

Figure 4-7. Steel pipe fittings are available to make virtually any connection or turn imaginable. (Courtesy of Midland College.)

Fiberglass Pipe. Fiberglass is a fiber-impregnated plastic material that is almost as strong on a pound-for-pound basis as most steels and is virtually corrosion-free. Yet fiberglass pipe is light and can be easily cut to any length. Fiberglass pipe is rapidly becoming a viable replacement for steel pipe in many flowline applications. Fiberglass pipe is sized by its outside diameter and is available in the same schedules as steel pipe.

Plastic Pipe. In the last few years a flexible plastic pipe has been developed for use in flowlines. This pipe comes on large reels of several hundred feet and is handled much like garden hose. Plastic pipe is generally limited to 125 psi but can be used in many low-pressure piping applications. Plastic pipe is available with outside diameters of 2–4 inches.

Asbestos Pipe. Asbestos (transite) pipe is compounded of a cementing material and asbestos fibers and is used for extremely corrosive environments. The asbestos allows the cement to expand and contract slightly without cracking, but major movement will crack the pipe easily. Transite predates plastic and fiberglass for use in corrosive environments.

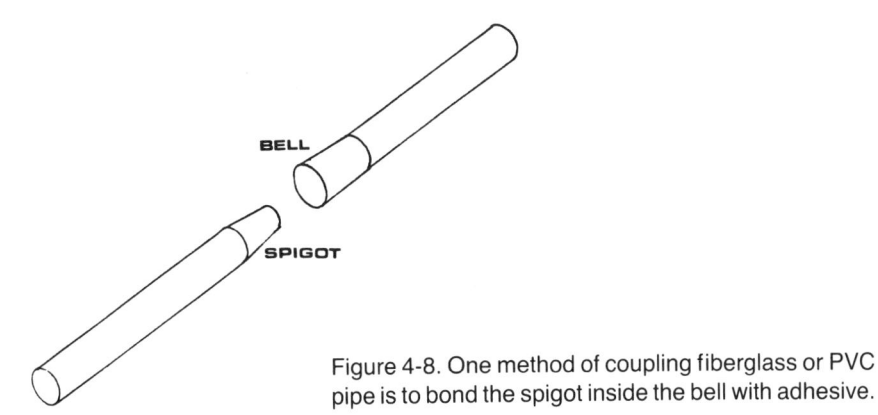

Figure 4-8. One method of coupling fiberglass or PVC pipe is to bond the spigot inside the bell with adhesive.

Cement-Lined Steel Pipe. Cement is poured into steel pipe then bored out. This leaves pipe with a cement sheath on its inside. This type of pipe is used for corrosive fluids, like oxygen-bearing water, to keep the fluids from contacting steel.

Coupling Methods

The method of coupling pieces of pipe differs with the type of pipe. Although another type of pipe may be used for long, straight sections, steel fittings are often used for terminating the pipe line. Each of the non-metallic pipe types have fittings to allow these materials to be joined with steel pipe at terminations.

Fiberglass and some plastic pipes (such as polyvinyl chloride—PVC) are formed with a bell on one end and a spigot or tapered end on the other as shown in Figure 4-8. The sections are joined and cemented with an adhesive. A special tapering tool is used to bevel the end of a piece of pipe that has been cut. All fittings for this type of pipe have a tapered shape just as the bell does. Some of the fittings for plastic and fiberglass pipe are shown in Figure 4-9.

Transite pipe is joined with collars using rubber or plastic O-rings. These collars are simply pressed in place and use neither threads nor adhesives. This sealing mechanism is not well suited for pressures exceeding about 50 psi.

There are two principal methods available for coupling steel pipe. One is to thread the two pipe ends and join them with a collar. Another method is to cut grooves in the ends of the pipes and join them with a grooved coupling (Figure 4-10). Fittings are available for both types of couplings.

122 Introduction to Petroleum Production

Figure 4-9. High- and low-pressure fittings are available to make many connections of fiberglass and plastic pipe.

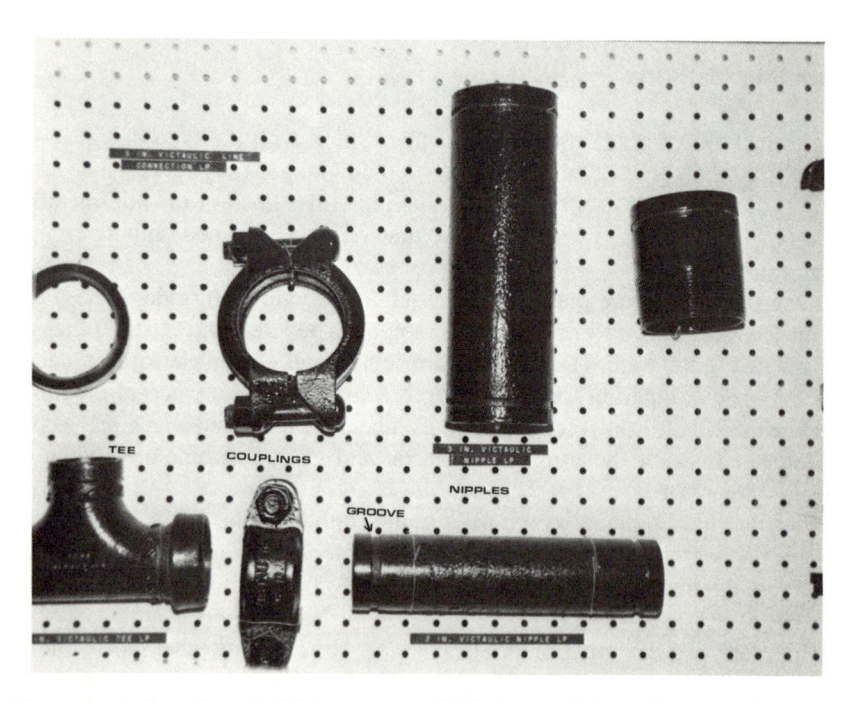

Figure 4-10. Couplings and fittings are available for steel pipe with grooved connections. (Courtesy of Midland College.)

Continuous flexible plastic pipe is joined by melting the ends of the pipe, pressing them together, and allowing the hot plastic to cool to form a continuous joint.

Flowline Installation and Maintenance

The pipe used for flowlines must be properly protected and handled as it is installed. Steel pipe should be stacked on sills to keep it off the ground. Plastic and fiberglass, on the other hand, are usually stacked directly on the ground because they are too flexible to rest on sills without excessive bending. The thread protectors and end caps should be left on the pipe to protect the ends and to prevent dirt and small animals from entering the pipe. As with other tubular goods, line pipe should not be walked on because the pipe can be damaged. Pipe should be picked up in slings to keep it straight.

When flowlines are installed, sharp turns should be avoided, since these accentuate water hammer. During installation, each joint of pipe should be checked for foreign material inside, the ends should be cleaned carefully, and the coupling installed.

Even the best flowline will eventually fail and leak. When leaks develop, flowlines must be replaced or repaired. Small leaks are sometimes repaired with flowline clamps pictured in Figure 4–11. For more serious leaks and for all leaks in non-metallic pipe, the entire section of leaking pipe is replaced.

Comparison of Flowline Materials

The selection of material used for flowlines is based on capital and installation cost as well as operating environment. Steel pipe is used more often than any other type. It is the strongest of piping materials, is available in a wide range of sizes, and fittings are available for most conceivable applications. The cost of purchasing, installing, and maintaining steel flowlines is generally high. Steel is easily attacked by corrosion from internal fluids as well as from water in the surrounding soil. One major advantage of steel pipe is that it can be located with metal detectors, and earthworking equipment can operate near steel pipe with little danger of damaging it. Another is that steel pipe can be used at any ambient temperature without losing its strength.

Plastic and fiberglass pipe can be used for almost any low-pressure flowline. They are inexpensive and easy to install. Neither plastic nor fiberglass can be located with metal detectors, and surface markers are needed to warn equipment operators of their presence. Plastic and fiberglass cannot be used when high-temperature fluids (more than 150–180°F) are being pumped with high pressure, since these materials soften at high temperature. The cost of maintaining plastic and fiberglass pipe is generally low.

Figure 4-11. Small leaks in flowlines may be temporarily repaired by placing a flowline clamp around the pipe and tightening the bolts until fluid flow stops. (Courtesy of City Pipe and Supply, Inc.)

Transite and cement-lined pipe are used for large-diameter gathering lines more often than for flowlines. In the past these pipes were used when large volumes of oxygen-bearing water (one of the more corrosive liquids involved in petroleum production) had to be moved, but they have been largely replaced by fiberglass and plastic. Both transite and cement-lined pipe can be damaged easily—transite pipe by crushing and cement-lined pipe by cracking and sloughing the cement sheath. Cement-lined pipe has additional disadvantages. A continuous cement layer between liquid and steel is very difficult to maintain, and there is no reliable method of joining the cement in two joints of adjacent pipe.

For the most part, it can be said that plastic or fiberglass flowlines may be used when the pipe may be buried deep enough to prevent mechanical damage, when the temperatures of internal liquid and surrounding earth are less than 150°F, and when many lines must be crossed because of the ease of coupling. Steel pipe should be used when high temperatures are expected, and when the flowline is to be installed on the surface.

High-Pressure Flowlines

Steel Pipe. Steel is usually used for high-pressure flowlines. The pipe schedule used depends on the internal pressure and the diameter of the pipe. Thick pipe is usually needed for large diameters to withstand the force exerted by high-pressure fluid. Most high-pressure steel flowlines use Schedule 40, 80, or 120 pipe.

Steel pipe used for high-pressure service is seldom joined with threaded couplings. Instead, steel is welded at all underground connections, and flanges are used for most connections above ground. When steel pipe is welded, the ends are beveled and several welding beads (single, continuous

Figure 4-12. Several weld beads stacked on each other are usually required for high-pressure welded connections. (Courtesy of Midessa Equipment Company.)

welds) are made as shown in Figure 4–12. A high-pressure flange is shown in Figure 4–13. Flanges are pressure-rated by an ANSI standard. This standard denotes not only the operating pressure but also the physical dimensions of the flange. A steel pipe may be plugged with a blind flange pictured in Figure 4–14.

Fiberglass Pipe. Fiberglass pipe has been improved so much in the last few years that it is now available for use when pressure exceeds 1000 psi. High-pressure fiberglass pipe uses a threaded-coupling method instead of adhesives and can be installed much more easily than steel. The pipe is less expensive than steel but has lower pressure and temperature limitations. A fiberglass-to-steel adapter must be used before any connections may be made because few high-pressure fiberglass connections and fittings are available.

Headers

A header in a gathering or distribution system provides a means of joining several flowlines into a single gathering line. Valves are sometimes provided on each pipeline entering or leaving the header so that lines can be isolated during normal operation or for maintenance.

Figure 4–15 shows the pipe and valve arrangement of a header that is commonly used for manually operated leases. All valves are easily visible

126 Introduction to Petroleum Production

Figure 4-13. Raised-face flanges are usually used for high-pressure pipe connections that may be disconnected occasionally. (Courtesy of Midessa Equipment Company.)

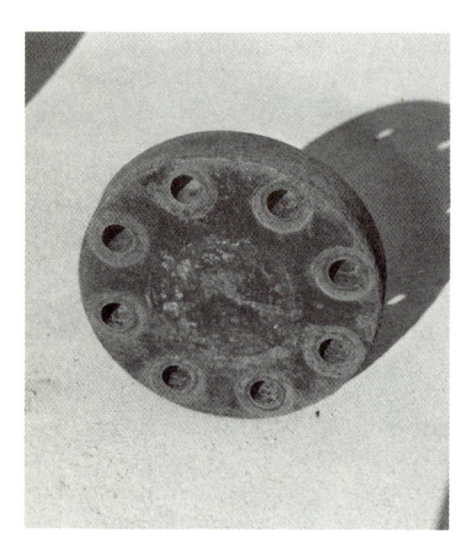

Figure 4-14. A blind flange has no opening and may be used to plug high-pressure pipe.

Surface Gathering Systems 127

Figure 4-15. This test header uses manually operated valves to select wells for testing.

and accessible from the top. This arrangement uses test and production headers. To route one well into the test header, the test valve is opened and the production valve closed. All other production valves are open and test valves closed.

Valves

Several types of valves are used for headers, but these same valves may be used for many other applications in petroleum production. The following describes most of the valves used in petroleum production and some applications of them.

A valve may be described as a device for limiting or blocking the flow of fluids. A valve may be designed for "on/off" operation (fluid is allowed to flow at full velocity or not at all), or a valve may be designed for throttling service (fluid velocity may be adjusted by partially opening or closing the valve). A choke is an example of a throttling valve.

Plug Valves. These valves (Figure 4–16) are used for on/off operation with low-pressure fluids. The steel plug is rotated to allow or prevent flow. The grease between the plug and the body keeps fluid from leaking around the closed valve, and regular lubrication is essential for proper operation. Plug valves are inexpensive and durable, but they are not well suited for throttling

128 Introduction to Petroleum Production

Figure 4-16. A plug valve has a steel plug which may be rotated in the body to allow or stop fluid flow. (Courtesy of City Pipe and Supply, Inc.)

Figure 4-17. The opening in a hollow ball may be rotated parallel to the direction of flow to allow flow and perpendicular to prevent flow in a ball valve. (Courtesy of Midland College.)

service and are used mostly for headers and for isolating various equipment. Plug valves are usually subject to corrosion and are not suitable for high temperature that would soften the grease. Because plug valves are inexpensive, they are often replaced rather than repaired.

Ball Valves. This valve (Figure 4–17) is similar to a plug valve, except that a valve seat is used to keep fluids from leaking around the ball. The ball is made of chrome steel for normal applications and stainless steel for highly corrosive environments. The seats are made of plastic or nylon material, but they may be special compounds or even stainless steel for high temperature or very corrosive duty. Although the ball valve was originally designed as an on/off valve to replace the plug valve, application has proved ball valves to be excellent throttling valves. When the ball or seats wear to the point that they do not seal properly, the ball and/or seats may be replaced easily. Lubrication is not required for most ball valves.

Gate Valves. This valve (Figure 4–18) is another type of on/off valve used primarily for liquids. To allow unrestricted flow, the gate must be raised completely. This valve may be used for throttling, but it is designed for on/off operation.

Needle Valves. Figure 4–19 shows a needle valve which uses a cone-shaped metal stem and matching seat to regulate fluid flow. A choke is a needle valve for high-pressure gas service. Needle valves are designed for throttling fluid flow and are not usually used for on/off operations.

Wafer Valves. Figure 4–20 shows a wafer or butterfly valve. The wafer is rotated against a seat to block flow and is turned parallel to the direction of flow for unrestricted flow. This valve is designed for on/off operation and does not perform well as a throttling valve.

Block and Bleed Valves. Some on/off valve applications are more affected by the possibility of leaks than others. Block and bleed valves (Figure 4–21) are used when it is imperative to know if a valve is leaking when closed. When the bleeder stopcock is opened, fluids will drain if the valve is leaking. This feature provides a positive indication of the valve's integrity. A double block and bleed valve (Figure 4–22) is used for applications when pressure on both sides of the valve is above atmospheric pressure. Again, the stopcock, which is left open when the valve is closed, indicates the condition of the valve's sealing mechanism.

Diverting Valves. These valves are used to route fluid through one of several possible paths. Figure 4–23 shows the operation of a three-way, two-position ball valve, while Figure 4–24 shows a three-way, three-position valve. Another diverting valve (Figure 4–25) uses a pneumatic operator to divert flow from one path to another. Figure 4–26 shows a valve section that is designed for use on production headers. One section is used for each flowline entering the header as shown in Figure 4–27. The hand-wheel operating assembly may be replaced with a pneumatic operator (Figure 4–28).

130 Introduction to Petroleum Production

Figure 4-18. A gate valve has a solid plate that may be lowered to block flow or raised to allow flow. (Courtesy of Midland College.)

Figure 4-19. A needle valve uses a cone-shaped plug and seat to block and throttle flow. (Courtesy of Midland College.)

Surface Gathering Systems 131

Figure 4-20. The wafer in a wafer valve may be rotated to allow or block fluid flow. (Courtesy of City Pipe and Supply, Inc.)

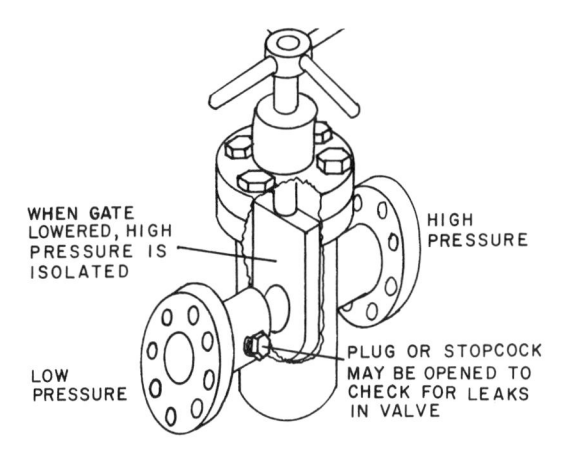

Figure 4-21. A block and bleed valve allows inspection to see that the valve is not leaking.

Check Valves. These are valves that allow unobstructed flow in one direction, but completely block flow from the opposite direction. Figure 4–29 shows a common check valve.

Valve Operators

Manual Operators. Every valve requires a method of moving the working parts. Manually operated valves simply use a handle or hand wheel such as would be found on a water hydrant. Valves which require only 90° rotation,

132 Introduction to Petroleum Production

BALL VALVE CLOSED

WHEN PLUG OR STOPCOCK OPEN LEAK FROM EITHER DIRECTION EVIDENT

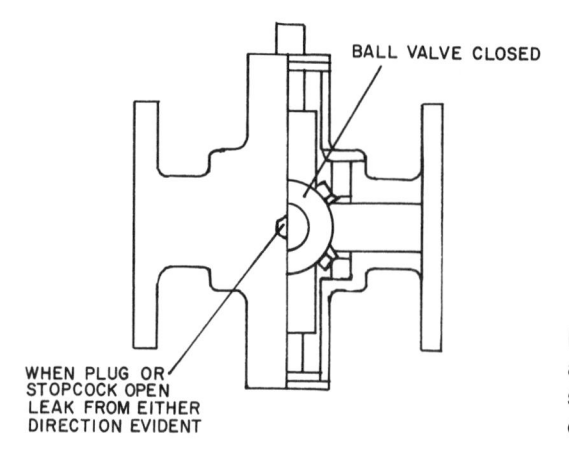

Figure 4-22. A double block and bleed valve allows inspection for a leak from either end of the valve.

MECHANICAL STOPS AND DETENTS ASSURE THAT VALVE CAN BE IN THESE POSITIONS ONLY

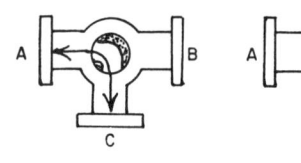

FLOW THROUGH A AND C

FLOW THROUGH B AND C

Figure 4-23. A three-way, two-position valve allows selecting one of two possible flow paths.

FLOW THROUGH A AND C   FLOW THROUGH B AND C

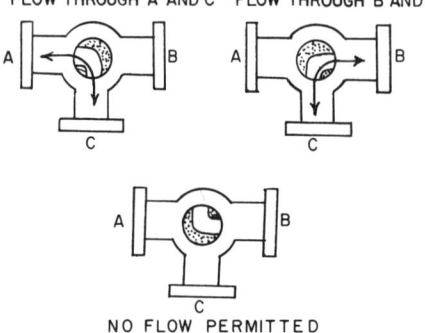

NO FLOW PERMITTED

Figure 4-24. A three-way, three-position valve allows selecting flow paths or stopping flow altogether.

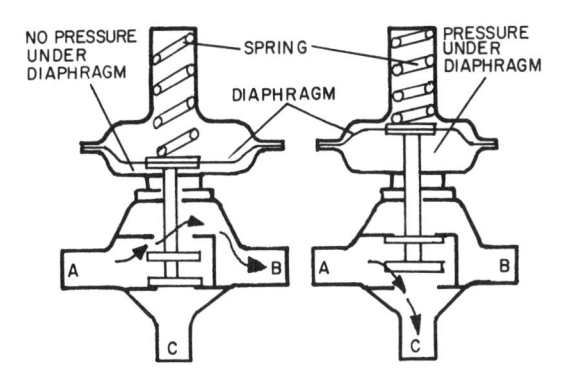

Surface Gathering Systems 133

Figure 4-25. A pneumatic diverting valve allows selecting flow paths by applying pressure to a diaphragm operator.

Figure 4-26. One type of header valve has plungers that act as valves and channels to act as pipe between valves to perform all header functions.

such as plug, ball, and wafer valves, use simple handles pictured in Figure 4–30. For many years, manual operators were the only ones available, and at present they are still the predominant form. Pneumatic and electric operators are used for remote or unattended operation.

Pneumatic Operators. These operators use a rubber or neoprene diaphragm to move valve stems. Gas pressure is applied to one side of the diaphragm, and a force equal to the product of the pressure and the surface area of the diaphragm is exerted against the stem. When the pressure is released, a spring returns the diaphragm to its original position as shown in Figure 4–31. Pneumatic operators are generally used for linear valves (those that move back and forth instead of rotating).

134 Introduction to Petroleum Production

Figure 4-27. Several header valve sections may be bolted together to form a complete header.

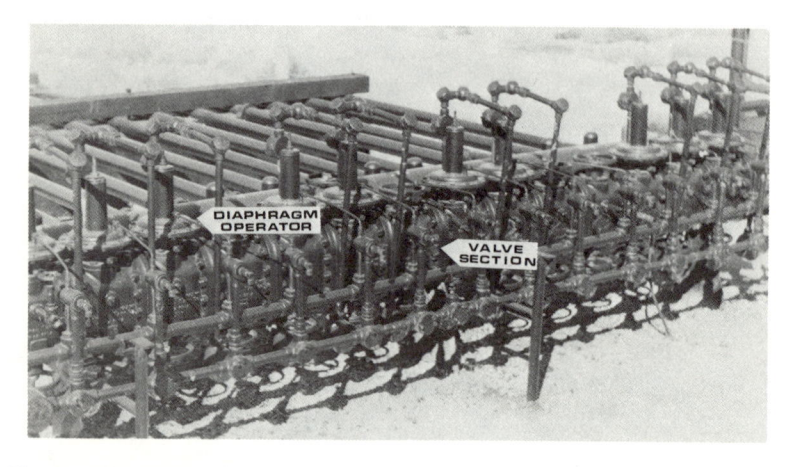

Figure 4-28. A test header may use valve sections operated pneumatically.

Electric Operators. This operator type uses a small electric motor of one-quarter horsepower or smaller and a gear train as shown in Figure 4–32. Electric operators are used with rotary valves or linear valves with a threaded stem. Some valves such as plug, ball, and wafer valves operate by moving 90° back and forth. These valves require an operator which can not only operate through an angle, but can also reverse directions. Figure 4–33 shows an electric actuator using a motor with two sets of windings (acting as if there are two motors each causing opposite rotation) for reversible operation and cams and limit switches to operate through any angle.

Surface Gathering Systems   135

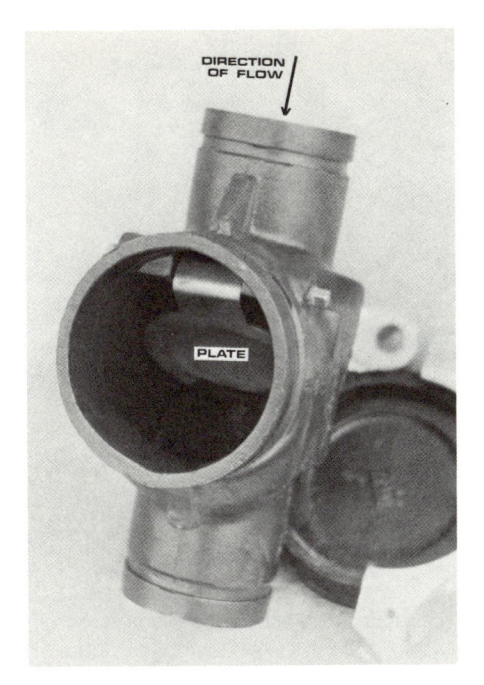

Figure 4-29. One type of check valve has a hinged plate that swings up to allow flow in one direction but swings down to prevent flow in the opposite direction. (Courtesy of City Pipe and Supply, Inc.)

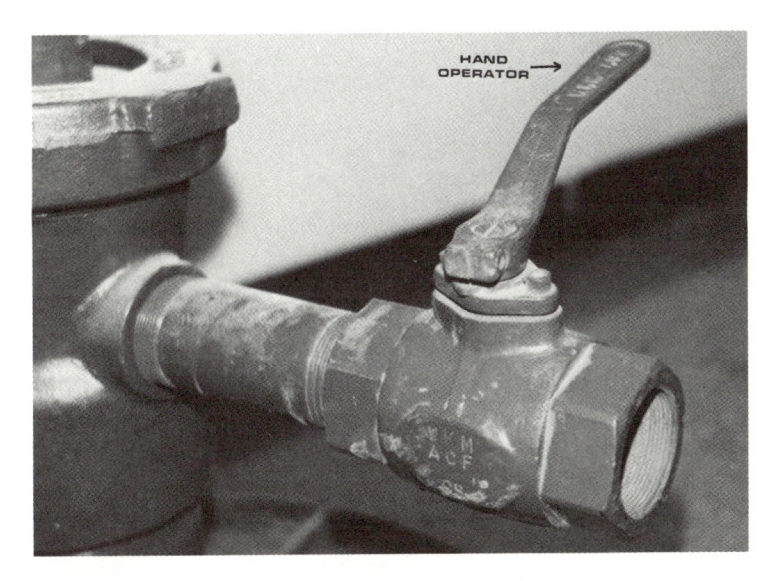

Figure 4-30. Most valves are equipped with handles for hand operation. (Courtesy of Midland College.)

136 Introduction to Petroleum Production

Figure 4-31. A diaphragm operator uses pressure and a spring against a diaphragm to operate other devices.

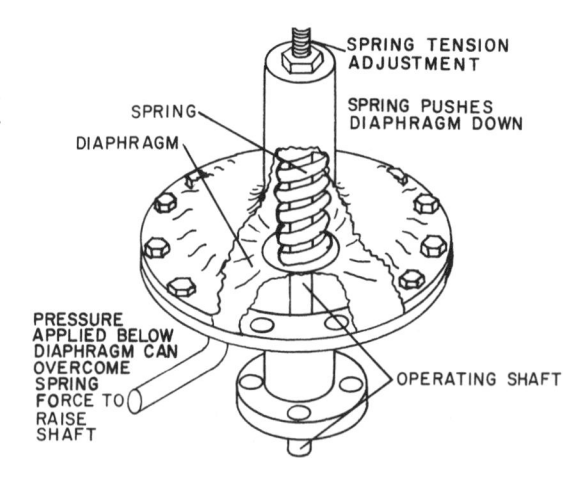

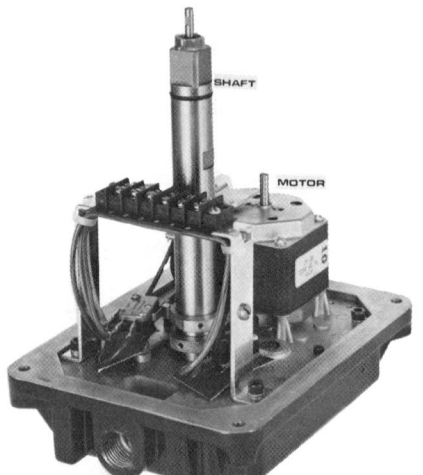

Figure 4-32. Some valves are equipped with actuators which have electric motor assemblies to rotate the valve in only one direction. (Courtesy of Raymond Control Systems.)

Figure 4-33. Other electric actuators provide for rotation in either direction and can provide rotation through a predetermined angle. (Courtesy of Raymond Control Systems.)

Valve Sizing and Specifications

The size of a valve is the size of the pipe to which the valve is connected. For example, a valve connected to one-inch diameter pipe is called a one-inch valve. Valves are available in sizes from one-eighth inch through 36 inches and larger.

Most valves smaller than two inches are connected to pipe using threaded couplings. The valves are usually supplied with a female thread at each end. Steel pipe with a male thread can then be joined at each end.

Larger valves are available for threaded, grooved, or flanged couplings. The coupling method must be specified when a valve is bought.

The working pressure is the maximum at which a valve may be safely operated without bursting. Valves are pressure tested to 1.5 times the working pressure, but they must never be operated at this pressure.

A standard working pressure for low-pressure valves is 125 psi. A great number of pipe fittings and piping systems have this rating. If there can be said to be a division between low pressure and high pressure, 125 psi is the division point. Valves are available with working pressure ratings of more than 125 psi, but they are usually considered to be high-pressure equipment.

Flow Behavior in Gathering Systems

As fluids pass through a valve or any other fitting, turbulence is generated. As flow continues downstream, it may return to laminar flow if the piping design allows.

Turbulence, as well as the serpentine path the fluid must follow, causes more friction than would be encountered in straight pipe. This friction is exhibited as a difference in pressure across the valve. As expected, friction and the resulting pressure drop increase as flow rate increases. For every fitting and valve, there is a ratio of the pressure drop in psi to the flow rate in gallons per minute (gpm). This ratio is called the Cv factor. The size or type of every valve and fitting should be determined based on this factor because the pressure drop for a given rate is less for a large valve or fitting (or a fitting that allows a gradual change in direction) than for smaller valves or fittings that cause sharp direction changes.

Cavitation is a phenomenon that can occur (mostly in valves but also in some fittings) when the pressure changes cause free gas to be liberated. The gas-liquid mixture can cause mechanical action much like water hammer that can batter the internals of a valve to the point of destruction. The gas released may be dissolved natural gas, light hydrocarbons, or water vapor boiled off the liquid by the sudden pressure reduction.

Cavitation is most serious in valves because its effect can be the same as sand blasting the machined surfaces of the working parts. Selecting a large enough valve to avoid a great pressure difference will usually prevent cavitation. However, even when the optimum valve size is selected, cavitation can still occur. In such cases valves equipped with special cages, flow channels, and baffles should be used to prevent cavitation. The initial cost of these valves is usually high, but this cost is less in the long term than continuously replacing the internals of valves.

Erosion is another phenomenon that can occur in any valve or fitting through which liquid flows. Just as water can erode soil and rock on the earth's surface, liquid flowing through a valve or fitting can erode the material of the fitting or valve. This erosion can cause pitting and distortion that eventually destroys the equipment.

Erosion is related to the velocity (in feet per second) of liquid flow. It is not necessarily affected by turbulence and is not directly reflected by a pressure difference. The only way to reduce erosion is to reduce the liquid's velocity. This can be done by selecting larger fittings and valves and by selecting equipment that does not use small orifices which cause high-velocity flow.

Gathering System Design

The selection of the type of gathering system is a general specification. Once the type has been decided, the details of the system must be specified. Such details include routing of flowlines and trunk lines, selection of valves and fittings, and determining the coupling methods and pipe to be used.

When determining the routing of flowlines, the effect of friction and turbulence should be considered. When possible, flowlines should be routed to have the fewest turns possible. When a turn is necessary, a fitting should be used that gives the most gradual turn possible.

Headers should also be designed to minimize friction and turbulence. The fittings and valves should be selected based on the effects of friction, cavitation, and erosion—not on cost alone. When possible, headers (or any other concentration of pipe and fittings that cause many direction changes) should be designed to have minimum turns and bends.

Fluids from producing wells are accumulated and transported by gathering systems. The end result of a gathering system is to move fluids to surface processing equipment. Since natural gas is almost always associated with petroleum production, some gas processing equipment is required for almost all leases. Gas handling equipment is discussed in Chapter 5.

Chapter 5
Gas Processing

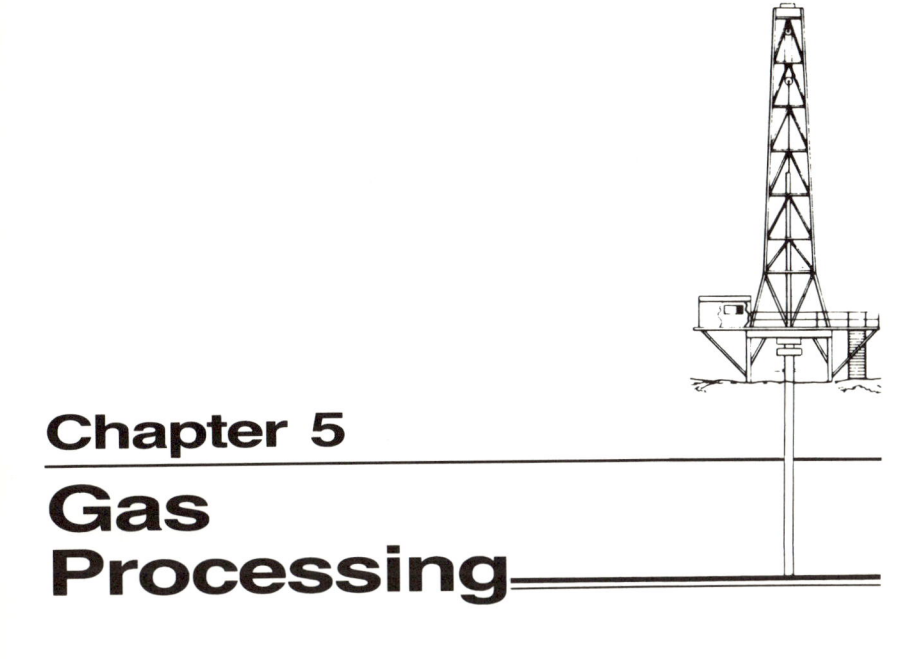

Gas processing is required to some extent on every petroleum producing lease. Even when gas is not the principal hydrocarbon, it can occur in free form because of pressure differences. On many small oil leases, gas is not produced in marketable volumes and must be disposed of by burning in flares. This is not a common practice, but it is necessary when there is no place to send gas. It is a common practice to flare gas when a process facility is stopped for emergencies or for maintenance. When normal processes are stopped and high pressures might develop, gas must be vented to the atmosphere and burned. These conditions occur infrequently and are short lived.

Gas heaters, which were discussed in Chapter 2, are used to prevent the formation of ice and hydrates at a well location or in the flowline. Heaters may also be used when liquids are produced in cold climates to prevent freezing or paraffin accumulations. Heaters are sometimes required by pipeline companies purchasing gas from a lease but are used on many leases simply as a matter of good operating practice.

Whether or not heaters are used, gas produced from wells must eventually reach surface processing facilities. The first stage of processing gas is to remove liquids from the gas. This is done in devices called two-phase separators.

140 Introduction to Petroleum Production

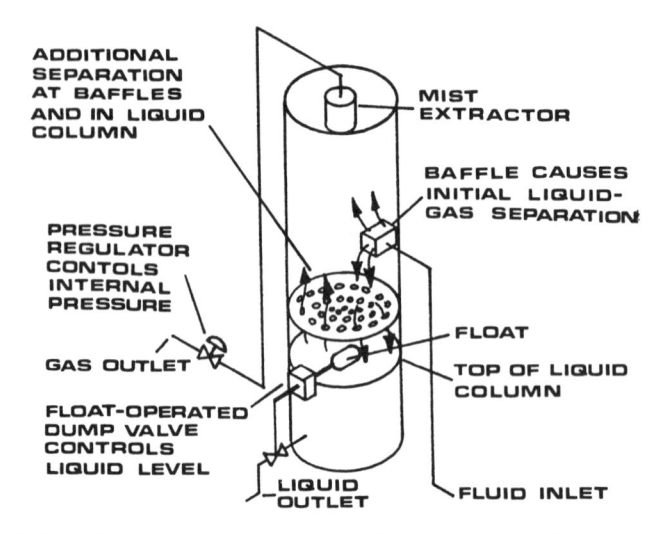

Figure 5-1. Liquid-gas separation is accomplished in a two-phase separator by agitation and settling action.

Two-Phase Separators

Two-phase separators are process vessels that segregate the liquid and gas phases and keep them apart until liquid leaves the vessel by one route while gas leaves by another. Separators are used to remove large volumes of liquid from gas.

There are several mechanisms that play a part in separation of liquid and gas. First, when the pressure of a fluid is reduced, less gas can be held in solution by the liquid. Some of the dissolved gas is liberated as free gas. Second, when fluid is allowed to stand for a time, minute gas bubbles will rise to the surface of the liquid and escape as free gas. Third, turbulent flow allows bubbles of gas to escape more rapidly than laminar flow, and many separators have parts that intentionally induce turbulence for this purpose. Several separator designs employ these mechanisms in different ways to achieve high-efficiency gas separation for different types of flow streams.

Vertical Separators

Figure 5–1 is a view of a vertical, two-phase separator. Fluid enters the vessel near the top and immediately strikes a box-like baffle. The baffle's purpose is to cause turbulence. Liquid containing some free-gas bubbles falls to the bottom of the vessel where it remains for a time; the length of time liquid stays in a separator is called the *retention time*. The retention time of a separator is determined by its internal volume which, in turn, is directly related to the inside diameter and height of the vessel. Liquid leaves the vessel at the bottom, while gas exits through the top. Vertical separators are most efficient for high flow rates, but they do not allow much retention time in comparison to other types of separators.

Vertical separators have a maximum operating pressure or working pressure just as valves and fittings do. Separators may be rated from 30 psi to more than 10,000 psi. The pressure that a separator can withstand is determined partially by construction techniques but mostly by the thickness of the walls and end caps of the vessel. High-pressure separators are generally tall, slender vessels, about 20–30 feet tall and 10–20 inches in diameter. Low-pressure separators are usually shorter vessels about 10–20 feet tall and 2–6 feet in diameter.

Many separators are equipped with a mist extractor at the top. This is an arrangement of baffles through which gas must pass before leaving the separator. Any droplets of liquid carried by the gas collect on these baffles and drip back into the separator. Thus, liquid droplets are not allowed to leave the separator with gas.

Liquid Level Control. The level at which liquid stands in the separator is controlled by the rate at which it is allowed to flow out of the separator. Liquid level controls use floats or level detectors to open the discharge valve when the liquid is above the desired depth and to close the valve when the liquid is below this depth. Figure 5–2 is a view of a simple, float-operated discharge valve assembly. When the float rises as liquid depth increases, the valve is opened, allowing fluid to leave the vessel. In this application the discharge valve (or dump valve as it is often called) throttles liquid flow. Sometimes this throttling action causes too much pressure drop or causes the separator to empty itself completely (an undesirable situation), and a snap-acting dump valve arrangement (Figure 5–3) is used to cause on/off operation. The float operates a pilot (a small pneumatic valve) which supplies gas pressure to the diaphragm of the dump valve. In either method of operation the level at which the float actuates the dump valve is adjusted by changing the length of the linkage. The dump valve is usually a wafer valve.

Another liquid level control method uses pneumatic control devices such as pictured in Figure 5–4. As liquid rises in the separator, supply gas is routed to a pneumatically operated valve configured for on/off or throttling operation. This arrangement is being used more and more because it is very dependable and requires little, if any, maintenance.

Separator Pressure Control. The pressure inside a separator is controlled by the rate at which gas is allowed to leave the separator. A back-pressure regulator is a diaphragm-operated valve that maintains a constant upstream pressure while allowing fluid to flow through it. Figure 5–5 shows a spring-

142 Introduction to Petroleum Production

Figure 5-2. A float-operated dump valve provides a simple method of removing liquid from a separator.

Figure 5-3. A snap-acting dump valve arrangement can be used for rapid operation of the valve without any throttling action.

Gas Processing 143

Figure 5-4. A pneumatic float-operated valve may be used for rapid movement as well as throttling action.

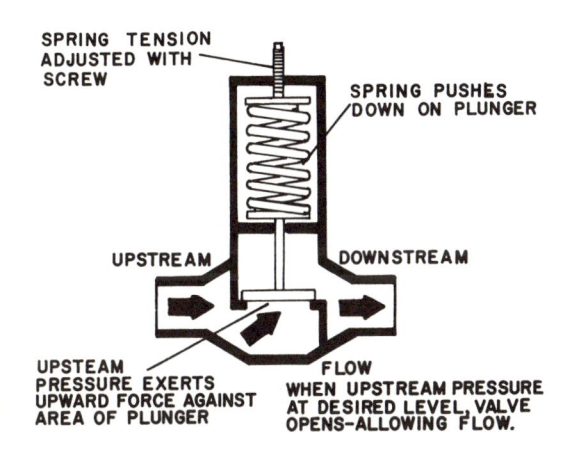

Figure 5-5. A back-pressure valve uses spring tension and pressure against a plunger to maintain constant pressure ahead of the valve.

operated, back-pressure valve in which the upstream pressure is regulated by adjusting the spring tension on the diaphragm. Figure 5-6 shows a pilot-operated, back-pressure valve in which spring tension in the pilot (Figure 5-7) regulates the upstream pressure. Pilot-operated valves are used for high-volume and high-pressure service because the tension in a large spring-operated valve may be too much to adjust manually; whereas, the tension in the pilot can be adjusted easily.

Separators are usually equipped with vent valves for safety purposes. Occasionally, the back-pressure valve or one of the fluid outlet lines plug, and the internal pressure in the separator can climb rapidly. Vent valves are used to allow gas to escape before the vessel bursts. Figures 5-8 and 5-9 show two types of vent valves. The outlet of the vent valve is piped to empty tanks

144 Introduction to Petroleum Production

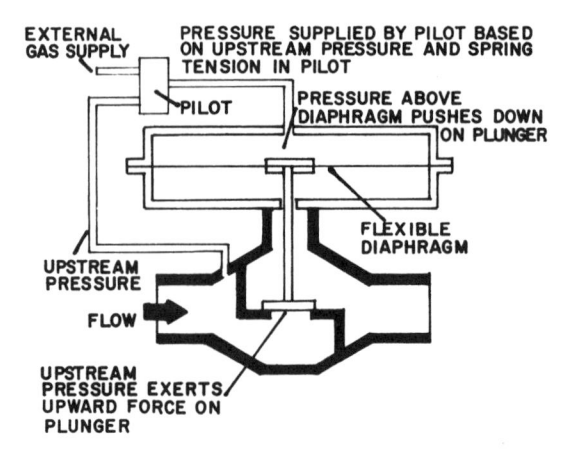

Figure 5-6. A pilot-operated diaphragm back-pressure valve uses spring tension in a pilot to apply pressure against a diaphragm to regulate pressure ahead of the valve.

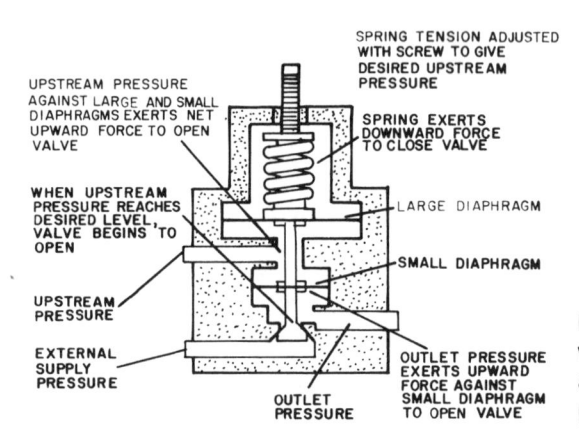

Figure 5-7. The pilot used with a pilot-operated valve is a small, spring-operated pressure regulator.

or to a flare where the escaping gas is burned until the internal pressure reaches normal conditions again. Vent valves should be very simple, durable, and dependable because they may be idle for months or even years (probably unattended) before being put into operation.

Rupture Disks. Many separators are provided with back up high-pressure safety devices called rupture disks. These are thin, concave, metal disks (Figure 5–10) installed in the top of the separator which are expected to rupture and release excessive pressure. The pressure at which a disk breaks is determined by its thickness and diameter. After the problem causing the high pressure has been corrected, a spent disk can be replaced easily.

Gas Processing   145

Figure 5-8. One type of relief valve operates when internal pressure exceeds the pressure of an adjustable weight against a plunger.

Figure 5-9. Another type of relief valve opens when vessel pressure exceeds the spring-tension setting in the valve. (Courtesy of Midessa Equipment Company.)

External Displays. Since most separators are usually inspected manually, some means must be provided to display internal operating conditions such as liquid level, temperature, and pressure. Internal liquid level in a low-pressure separator is shown with a sight glass, like the one in Figure 5–11, while level in a high-pressure vessel is indicated with a sight glass as in Figure 5–12. When it is necessary to display depth over a greater range than the length of the sight glass, two or more such glasses are used (Figure 5–13).

Internal temperature is shown with a thermometer as shown in Figure 5–14. Inside the tube of a thermometer (Figure 5–15) is a spiral tube. As

146 Introduction to Petroleum Production

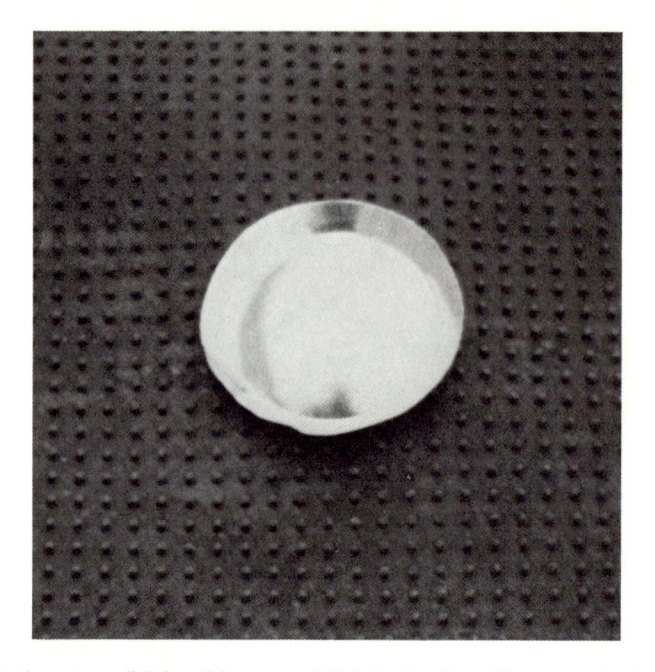

Figure 5-10. A rupture disk is a thin, curved disk that is placed in a vessel to burst and release pressure if too much pressure develops in the vessel.

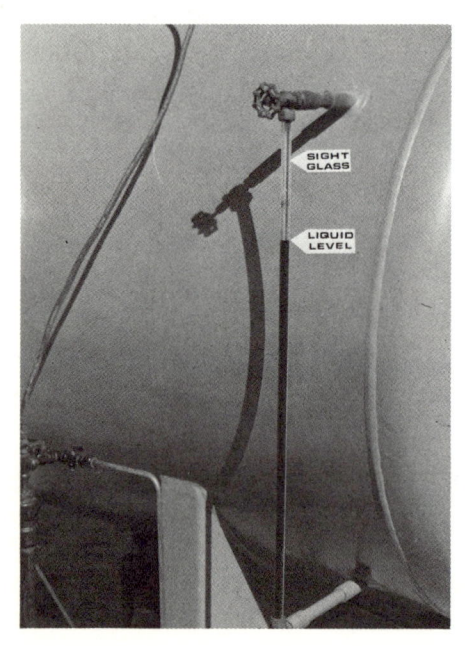

Figure 5-11. A sight glass allows internal liquid depth to be viewed without opening a vessel.

Gas Processing 147

Figure 5-12. A special type of sight glass is required to observe internal liquid depth in high-pressure vessels.

Figure 5-13. Deep or multi-phase liquid levels may be determined by using several sight glasses.

Figure 5-14. A thermometer indicates the temperature inside vessels or pipes.

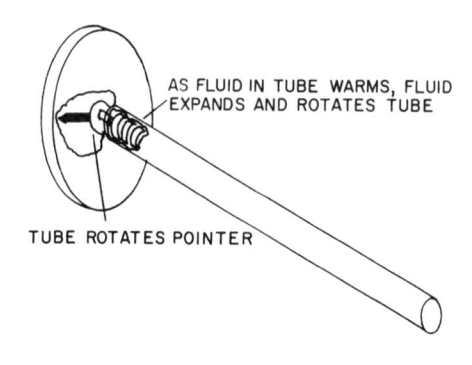

Figure 5-15. Some temperature gauges operate when fluid in a spiral tube expands and turns the tube and pointer.

temperature increases, the fluid in the tube expands, causing the spiral to rotate and move the pointer. Thermometers are installed in wells so that they are not directly exposed to the sometimes corrosive and high-pressure fluids inside a separator (Figure 5–16).

Pressure Gauges. There are many types of pressure gauges for use in separators. The oil-filled gauge depicted in Figure 5–17 and detailed in Figure 5–18 is a commonly used gauge for simple pressure measurement. This pressure gauge is inexpensive but is not particularly durable when exposed to the vibration, hostile environment, and corrosive atmosphere usually associated with separators. Pressure gauges like those in Figure 5–19 are much more durable and accurate, but they are also more expensive.

Horizontal Separators

A horizontal separator (shown in Figure 5–20) is used when long retention time is required to separate gas from liquid. Incoming fluid enters through a

Gas Processing 149

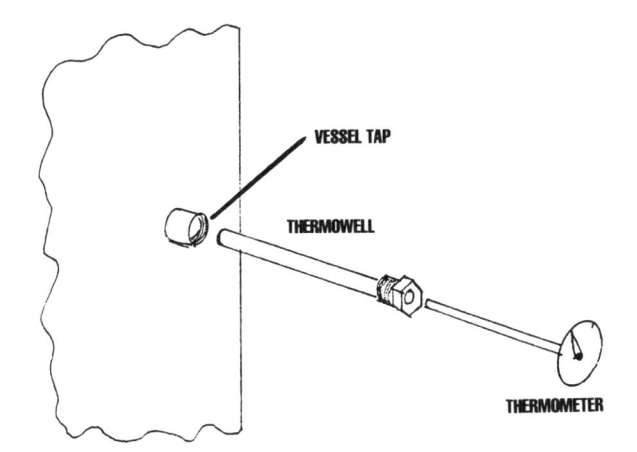

Figure 5-16. Thermowells are used so that delicate thermometers are not directly exposed to high pressure or corrosive fluids in vessels.

Figure 5-17. Oil-filled pressure gauges are durable enough to withstand poor environmental and operational conditions and still provide adequate pressure measurements.

baffle that causes turbulence, and the liquid falls into the bottom part of the separator where liquid is allowed to stand for some time. The retention time is determined partly by inlet flow rate and partly by the size of the vessel. Long retention time gives gas bubbles time to rise out of the liquid. A mist extractor prevents liquid droplets from leaving the separator with the gas.

The liquid level and pressure-control devices are the same as those used for vertical separators. Vent valves and rupture disks are also required on some horizontal separators.

150 Introduction to Petroleum Production

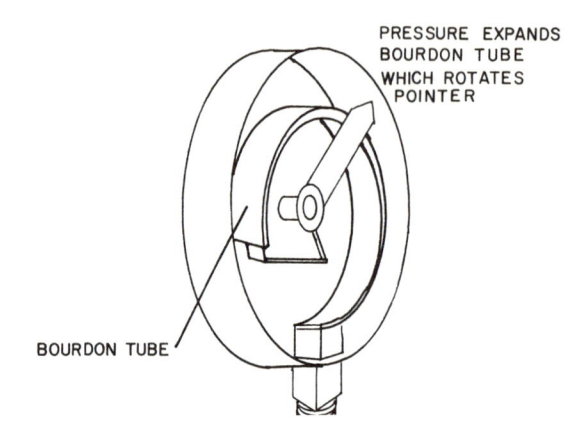

Figure 5-18. A pressure gauge contains a tube which expands and rotates a pointer to indicate increasing pressure.

Figure 5-19. Very accurate gauges are available to measure pressure as well as vacuum and differential pressure. (Courtesy of City Pipe and Supply, Inc.)

Gas Processing 151

Figure 5-20. Horizontal separators provide long retention time for applications when gas is difficult to separate from oil.

Figure 5-21. Horizontal separators are used when space is at a premium and are commonly found on skid-mounted assemblies. (Courtesy of Engelman-General, Inc.)

Horizontal separators are compact and well suited for skid-mounted assemblies such as the one in Figure 5-21. A horizontal separator must be larger than the vertical separator that could process the same volume of fluid because the retention time is so much greater. In general, vertical separators are used for high-volume separation, while horizontal separators are used for slow gas separation. The decision to use one or the other depends on costs, operating environment, and fluid characteristics (like specific gravity and viscosity).

Other Separator Types

Spherical separators (Figure 5-22) are seldom used but deserve mention. Incoming fluid is forced against the top of the sphere and allowed to run down the sides of the sphere. This flow path causes turbulence. Retention time for this separator is usually long. This separator design is less efficient than either vertical or horizontal separators, but its compact shape is desirable in some crowded locations.

Scrubbers (discussed in Chapter 2) are small separators. They are usually used to remove small amounts of liquid from gas that is to be used to operate pneumatic control devices and instruments.

Some large gas processing facilities use inlet scrubbers (Figure 5-23) to remove small volumes of liquid from gas streams. These scrubbers usually act as a final separation stage before compression or dehydration because they are not intended to remove appreciable volumes of liquid. Inlet scrubbers are designed to capture a slug of liquid left in a gas stream because of malfunction of another separator.

Dehydrators

Dehydrators remove water and water vapor from gas before the gas is processed further. Water vapor has two undesirable effects on gas. First, it interferes with sweetening and refining processes. Second, water vapor can be responsible for the formation of hydrates.

Glycol Dehydrators

Water may be removed from natural gas using a glycol dehydration process diagrammed in Figure 5-24. Glycol is an organic compound in which water will dissolve. Gas is forced to bubble through the glycol contactor tower where the gas, water vapor, and glycol are forced into intimate contact. Water is absorbed by the glycol, while gas flows out the top of the contactor.

Glycol can absorb only so much water before becoming totally saturated. Saturated glycol is pumped to a gas-fired boiler where it is heated. Water,

Gas Processing 153

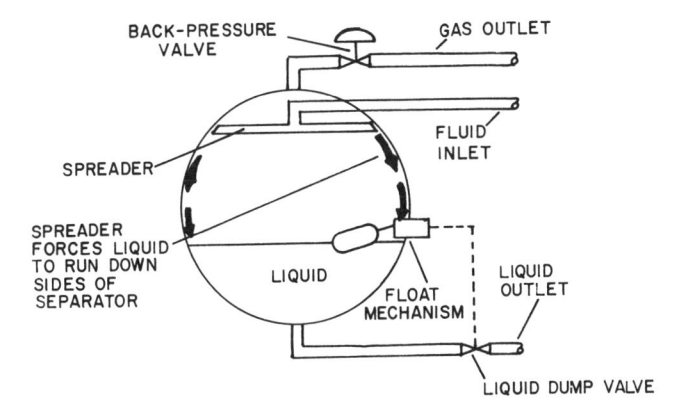

Figure 5-22. Spherical separators remove gas from liquid by exposing the liquid to the surface area of the sides of the vessel and allowing long retention time.

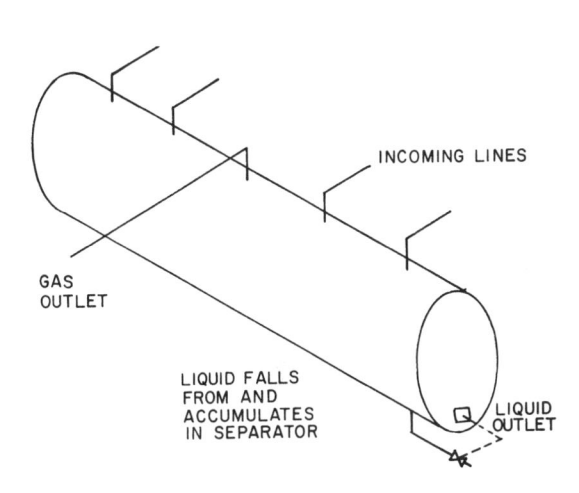

Figure 5-23. Gas processing facilities use an inlet separator to trap liquid before it reaches the facility.

which has a lower boiling point than glycol, is boiled out, and the water vapor is condensed into a tank where it may be removed with a pump. The regenerated glycol is cooled and pumped back into the contactor to attract additional water.

Figure 5–25 shows a small glycol dehydrator used to process instrument air in a large process facility. Larger versions of this unit are used to process produced gas in field gas process facilities, gasoline plants, and refineries.

The contactor and the glycol pump are subject to the pressure of the incoming gas. Therefore, these parts are constructed for high-pressure service when necessary. The boiler may be a low- or high-pressure vessel—depending on

154 Introduction to Petroleum Production

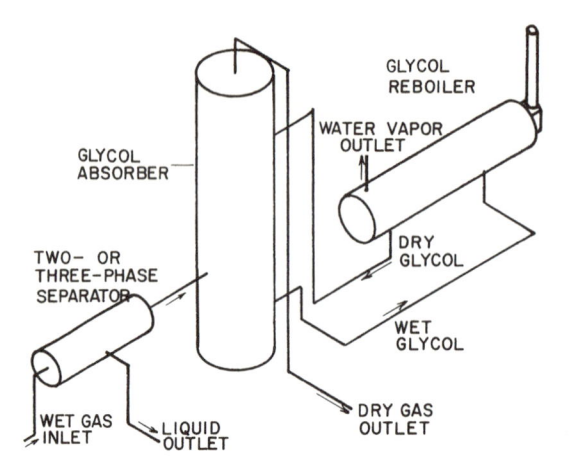

Figure 5-24. A liquid dessicant dehydrator uses glycol to absorb water vapor from gas.

Figure 5-25. A glycol dehydrator removes water from gas or air streams in a production operation. (Courtesy of Midessa Equipment Company.)

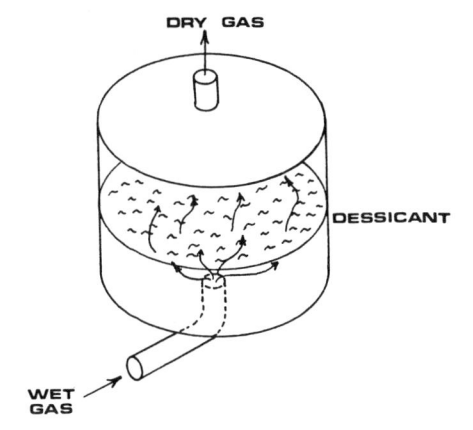

Figure 5-26. Gas containing water vapor may be passed through a solid dessicant dehydrator to remove the water.

the system design—but all boilers are constructed for moderate high-pressure simply because of the pressures involved in the processes of boiling and distilling.

Solid Dessicant Dehydrators

Another method of removing water vapor from gas is to use solid dessicants (a solid, granular medium which adsorbs liquids). A solid dessicant dehydrator is shown in Figure 5–26. Gas flows through a bed of solid material which adsorbs the water. The gas flows out through the top of the vessel after passing through the bed.

There is a limit to the amount of water the dessicant will adsorb. As this saturation limit is neared, the dessicant does not remove all water vapor, and some passes out of the dehydrator with the gas. Before the saturation point is reached, the dessicant must be regenerated. Water cannot be removed from some dessicants, and the material must be replaced periodically. Other dessicants give up water when they are heated. Some dehydrators have several dessicant beds as shown in Figure 5–27. Gas is routed through one or more beds until they become saturated with water. Then the gas is routed to other beds, and hot gas is passed through the saturated beds to remove water. Many dehydrators have several beds and pneumatically or electrically operated valves which switch gas and turn on burners automatically at regular intervals.

Solid dessicant dehydrators are also designed for vapor inlet. Liquid entering the beds can immediately saturate them; but unlike glycol dehydrators,

156 Introduction to Petroleum Production

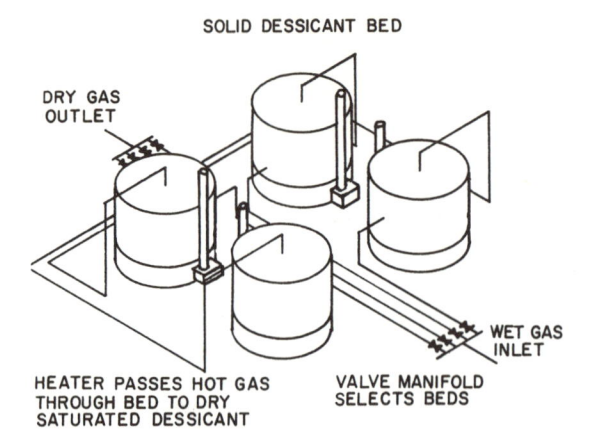

Figure 5-27. Multiple-bed dehydrators use several beds for dehydration while other beds are being regenerated with hot gas from gas-fired heaters.

solid dessicant beds are rendered useless by liquids and must be replaced entirely. Again, it is essential that liquid be removed entirely from the gas stream prior to entry into the dehydrator.

Instrument Gas Dehydrators

Pneumatically operated controllers and instruments often require supply gas or air that is completely free of liquids and water vapor. Small scrubbers (discussed in Chapter 2) remove liquids but not water vapor. Small, solid dessicant dehydrators, such as the one shown in Figure 5-28, are used to remove the water vapor. The dessicant is contained in a removable fabric sack which is replaced when the dessicant is saturated.

Another method of cleaning pneumatic supply gas is to use a small glycol scrubber shown in Figure 5-29. The scrubber is filled with glycol, and supply gas is bubbled through it. The glycol removes water vapor from the gas, and the water settles to the bottom of the scrubber where it can be drained periodically. This device also acts as a scrubber because, if liquid enters, it stops. Oil usually mixes with the glycol, while water settles to the bottom of the scrubber. The glycol in this small scrubber must be replaced periodically because no regeneration facilities are involved. Some service companies offer to collect glycol from these scrubbers, so it is not necessary to purchase new glycol each time the scrubber is serviced.

Gas Processing   157

Figure 5-28. A small, solid-dessicant dehydrator removes water from gas or air used to operate instrument and controls.

Figure 5-29. A glycol scrubber can also be used to dry air or gas as well as capture slugs of liquid.

Gas Compressors

Once gas has reached process facilities and its pressure is reduced in separators or other vessels, the pressure of the gas must be raised again to move it from one place to another. Gas is not pumped in the same ways as liquid but is moved only by raising its pressure significantly above the pressure at its destination.

Compressors are machines that raise the pressure of gas for transportation to another place. Sometimes compressors are needed merely to move gas within a surface process facility, but they are often used to transport the gas from a well to a natural gas pipeline for transportation across great distances.

All compressors work on the physical principle of the gas law. This law states that if the volume of a given quantity of gas is reduced, both the pressure and temperature of the gas are increased. Therefore, compressors work by reducing the volume of a gas and causing a pressure and temperature increase.

Types of Gas Compressors

There are only two compressor types used to any real extent in the production of petroleum: (1) positive displacement compressors, and (2) centrifugal (rotary) compressors. These compressors differ in the method, but not the result, of reducing volume to raise pressure.

Positive Displacement Compressors. A positive displacement compressor uses a piston in a cylinder through which gas flows. Inlet and discharge valves control the times when gas flows.

Figure 5–30 shows the operation of a positive displacement compressor. On the inlet stroke, gas enters the cylinder through the inlet valve as the piston is withdrawn from the cylinder. At the end of this stroke, the volume of the cylinder is filled with gas. On the compression stroke, the piston moves into the cylinder, reducing the volume and increasing the pressure of the gas. Near the end of the compression stroke, the discharge valve opens, and the high-pressure gas flows into the discharge header.

The rate at which gas flows through a cylinder depends on the size of the cylinder, the length of the piston stroke, and the speed of the crankshaft. One cylinder volume is displaced for each revolution of the crankshaft.

The volume of a cylinder ranges from several cubic inches for a small compressor, such as might be used in a service station, to several cubic feet for large pipeline compressors. The volume flow rate is determined by the crankshaft speed. If the shaft speed is 1200 rpm, 1200 times the cylinder volume flows per minute.

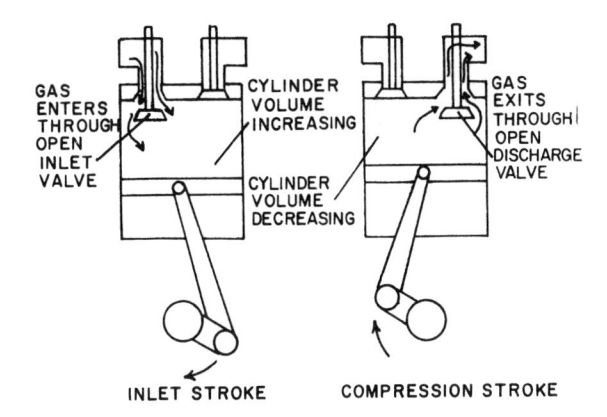

Figure 5-30. A positive-displacement compressor uses valves to route gas in and out of cylinders in which pistons compress gas by reducing its volume.

The pressure increase is directly related to volume reduction in a compressor cylinder. If the gas volume is reduced by a factor of 10 by the piston's movement, the absolute pressure is increased by a factor of 10. *Absolute pressure* is the gauge pressure plus atmospheric pressure. For example, the absolute pressure of 10 psig (pounds per square inch gauge) is 24.3 psi (psi absolute), since atmospheric pressure is about 14.3 psi. Most positive displacement compressors can increase pressure by a factor of 10–12. Mechanical constraints make a compression ratio of more than 12:1 impractical.

The actual discharge pressure of a positive displacement compressor is related not only to the compression ratio but also to the suction (inlet) pressure. Gas entering the cylinder does so at an inlet pressure that is not necessarily low. The discharge pressure of a single compression stage (accomplished by one or more cycliners acting in parallel) is the product of the suction pressure and the compression ratio. If gas at atmospheric pressure enters a 10:1 compression stage, the pressure is raised from 14.3 psia to 143 psia.

When the required compression ratio is more than 10–12:1, multiple-stage compression is required. This procedure is shown in Figure 5–31. Gas is compressed in the first stage, and the discharge of the first stage is the suction of the next stage. Using the previous example, if gas at atmospheric pressure enters a two-stage compressor with compression ratios of 10:1 in each stage, the suction pressure is 14.3 psia, the pressure between stages is 143 psia, and the discharge pressure is 1430 psia.

160 Introduction to Petroleum Production

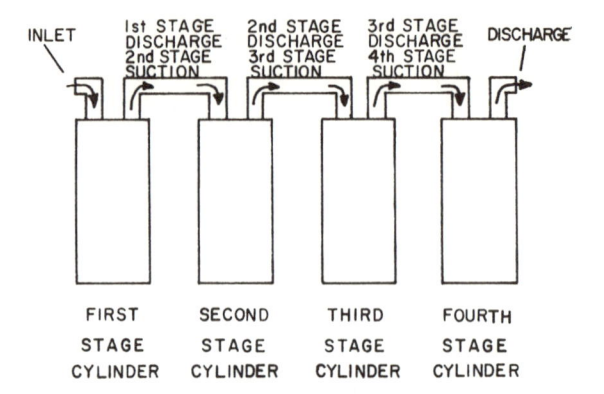

Figure 5-31. Staged compressors use several compression cylinders to sequentially raise gas pressure.

Figure 5-32. Multiple-stage compressors transport large volumes of gas at high pressure.

Figure 5–32 is a photograph of a multiple-stage compressor. The suction piping is very large in comparison to discharge piping because the actual volume at the discharge is much smaller (by a factor equal to the compression ratio) than at the suction.

Unlike liquid, the volume of gas pressurized and ready for transportation is not the same as it was at the lower suction pressure. To measure the volume of compressed gas, some standard volume measurement must be used.

As expected, a glycol dehydrator must employ several control and safety devices. The glycol levels in the contactor and boiler are supposed to remain constant, and liquid level control devices are required. The boiler and tank pressures are controlled with pressure regulators and/or back-pressure valves. Vent valves and rupture disks are often used to prevent excessive pressures in the vessels. Finally, pressure, temperature, and level gauges are required to describe the internal conditions in the dehydrator adequately.

Glycol dehydrators are designed to operate with gas and water vapor only. If liquid water or oil enters the dehydrator, its operation can be completely disrupted. The glycol will absorb and be contaminated by the oil and water. Most of the water and some of the oil can eventually be removed in the boiler, but some of the heavy hydrocarbons cannot be removed. The glycol must be removed from the unit and replaced in an operation that requires stopping the gas flow (and sometimes all fluid production on a lease) for several hours. To avoid this time-consuming and often costly maintenance, it is essential that incoming gas has already passed through separators and/or scrubbers before arriving at the dehydration equipment.

Gas volume is measured in standard cubic feet, which means the volume the gas would occupy if its pressure were 14.3 psia (sea level atmospheric pressure) and its temperature were 60°F. To convert the volume of compressed gas to its equivalent volume in standard cubic feet, the volume must be multiplied by the ratio of the absolute pressure to 14.3 psia, and by the ratio of 520 degrees to its present temperature plus 460 degrees. This arrangement expresses temperature on the Rankine or absolute scale.

In most measurements concerning petroleum, gas volumes are often measured in thousands or millions of cubic feet. To simplify the units of measurement, abbreviations are commonly used. The abbreviation scf means standard cubic feet. The prefix M is added to mean thousand cubic feet. 100 Mscf means 100,000 standard cubic feet. The prefix MM is used to mean millions: 10 MMscf means 10,000,000 standard cubic feet.

Centrifugal Compressors. Centrifugal compressors use a turbine-like impeller to compress gas as shown in Figure 5–33. Actually, the impeller increases the velocity of the gas, and the gas is compressed as its momentum carries it into the reduced volume of the discharge piping. If there is no open path through which gas may flow, the impeller simply spins, and the gas slips by the blades without being compressed. Each centrifugal compressor has an upper pressure limit at its discharge, and any attempt to attain higher pressure results in gas slipping by the impeller.

An actual centrifugal compressor (Figure 5–34) has a compression ratio of 5–10:1. Centrifugal compressors cannot generate the high discharge pres-

162 Introduction to Petroleum Production

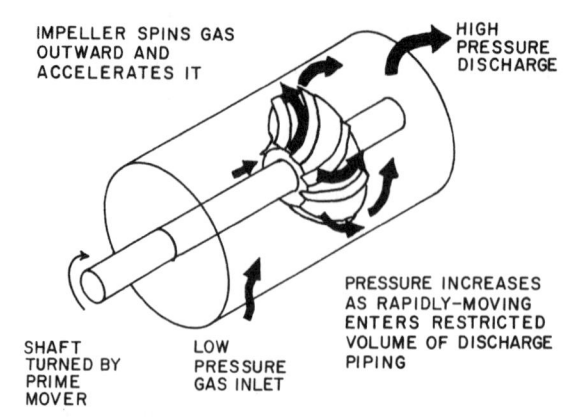

Figure 5-33. A centrifugal compressor raises gas pressure by using an impeller to accelerate gas into discharge piping.

Figure 5-34. Centrifugal compressors move small volumes of air or gas at low pressure.

sures possible with positive displacement compressors but are adequate for many low-pressure and moderate-volume applications.

Comparison of Compressors

The positive displacement compressor is the common compressor type in use today. It is limited to a compression ratio of 10–12:1, but its discharge pressure and volumetric capacity are limited only by the number of stages, cylinder size, and piping strength. Positive displacement compressors usually need a suction pressure of 30 psig or more; however, multiple-stage compressors are available which can have suction pressures at or less than zero psig (vacuum compressors). With this compressor, the first stage or two have compression ratios less than 5:1 so that they can handle the low pressure.

Positive displacement compressors are usually large, heavy units. They are usually very expensive, but can often be leased instead of purchased.

Centrifugal compressors are usually small, light, and relatively inexpensive machines. They can seldom function with high discharge pressure (several thousand psi). Compression ratios usually range from 5–10:1. It is possible to use centrifugal compressors in multiple-stage applications.

Prime Movers

Gas Engines. Gas engines are often used with compressors because the fuel source is so conveniently located. Small gas turbines are used for starter motors, but electric starters are also used. Either high- or low-speed engines may be used. Gas engines and compressors are often supplied as a skid-mounted package (Figure 5–35).

Gas Turbine Engines. A gas turbine engine, like the one shown in Figure 5–36, is a high-efficiency method of obtaining horsepower. Fuel gas is mixed with air and compressed. Electrically operated glow plugs ignite the fuel which expands as it burns. The rapidly expanding, hot exhaust gas rushes past an impeller on its way to the exhaust stack. The impeller spins and delivers torque to the output shaft.

Gas turbine engines have automated electrical and pneumatic systems for starting, control, and safety. Although gas turbine engines are complex, they are one of the most efficient methods of driving large loads of 1000 horsepower or more.

Electric Motors. Electric motors are the most cost efficient of prime movers for less than 1000-horsepower loads—if electric power is available at the

164 Introduction to Petroleum Production

Figure 5-35. Engine-compressor sets are often supplied as packages with the engine, coupling, and compressor already connected. (Courtesy of Compressor Systems, Inc.)

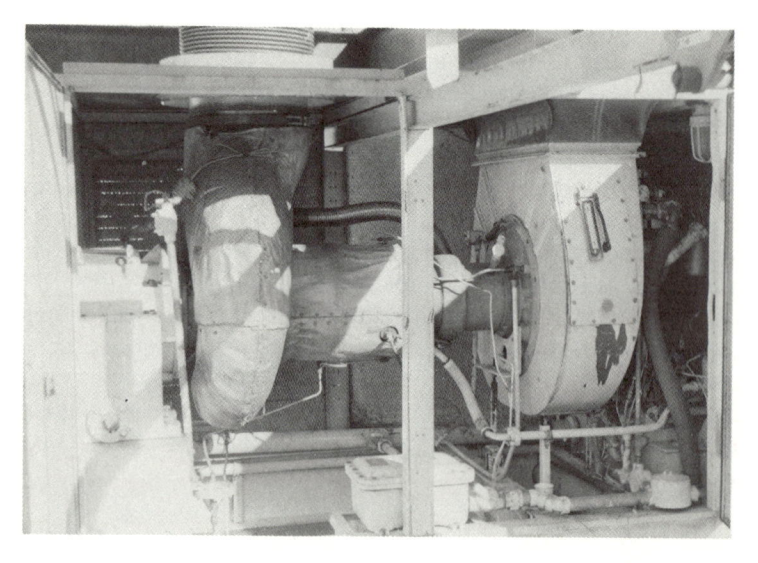

Figure 5-36. Compressors may also be powered by large, high-efficiency gas turbine engines.

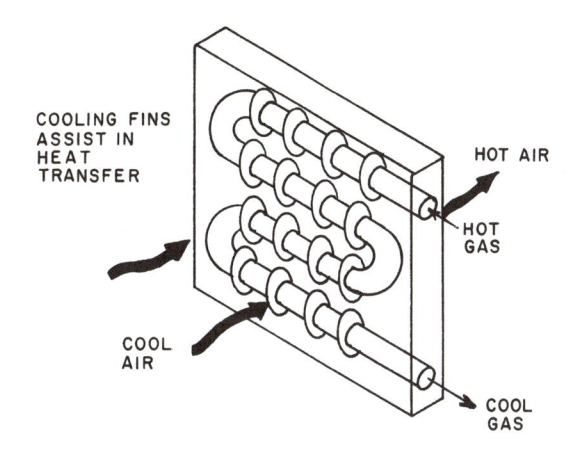

Figure 5-37. Gas coolers use cooling tubes and fins to transfer heat from hot gas to cooler air.

site. These motors are light, inexpensive, durable, and dependable. A disadvantage of electric motors is that their speed is fixed and cannot be easily varied.

Coolers

When a gas is compressed, its temperature and pressure increase. Although there is no danger of explosion, the hot gas is usually cooled before anything else is done with it.

Hot compressed gas is usually cooled with fans and coils like those diagrammed in Figure 5–37. Heat moves from the gas into the cooling fins, where it is carried to the atmosphere by the air driven by the fan. The temperature of high-pressure gas can be reduced drastically, but it is still above ambient temperature. Figure 5–38 is a photograph of a cooler.

Ancillary Equipment

Suction Scrubber Bottles. Liquid is incompressible, and any liquid passing through a compressor can seriously damage the machine. Most compressors have scrubbers upstream of their suction valves. These scrubbers are intended to trap liquids before they reach the compressor.

Engine and Compressor Controls. Every compressor has certain control and safety functions that must be displayed. Many compressors are equipped to stop running immediately if critical parts, such as the crankshaft bearings and the cylinders, get too hot. The compressor may be stopped if the suction pressure is too low or the discharge pressure too high. Suction, discharge,

166 Introduction to Petroleum Production

Figure 5-38. Hot gases are cooled as they pass through coolers which blow cool air over the tube bundles. (Courtesy of Compressor Systems, Inc.)

and lubricating oil pressures are displayed on pressure gauages. Engine information, such as lube oil pressure and temperature, cooling water temperature, and fuel gas pressure, are also displayed and used as shutdown parameters. Most of these critical data are handled by instruments, switches, and gauges on a control panel such as the one in Figure 5–39.

Compressor By-pass. Most compressors are designed to operate with a fixed suction pressure so the machine will furnish the desired discharge pressure. In some applications, however, the actual suction pressure is too low for the compressor to deliver the correct pressure. In such cases a by-pass arrangement, such as the one shown in Figure 5–40, is used to route a portion of the discharge gas back to the suction. This arrangement allows the suction pressure to be held at a fixed level. The use of a by-pass decreases the efficiency somewhat, but this small loss is more than offset by the improved operation with fixed suction pressure.

Skid-Mounted Compressors

In the last few years manufacturers have been supplying their compressor, prime mover, cooler, and all ancillary equipment in a single, skid-mounted package such as the one shown in Figure 5–41. This arrangement is much easier to install than the individual pieces of equipment, because the end user is only required to make suction, discharge, fuel, and electric connections.

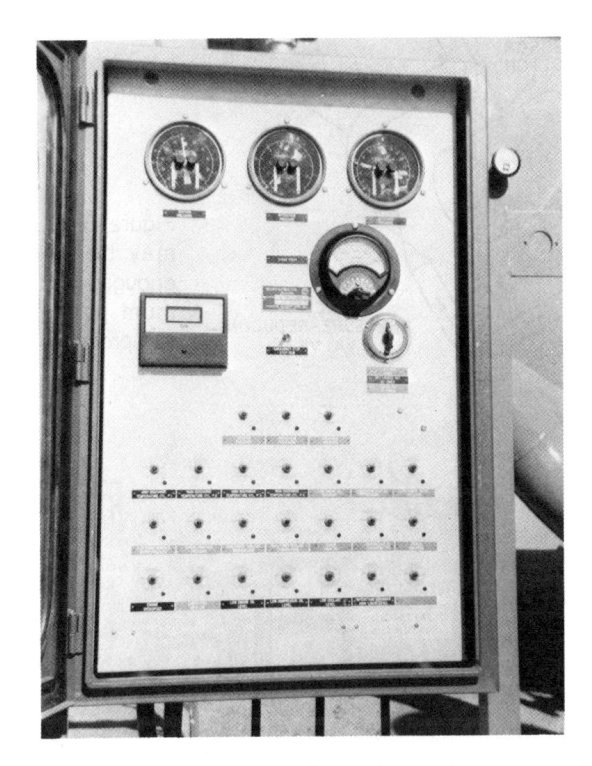

Figure 5-39. Operational information about the engine and compressor is displayed on a control panel which allows automatic and manual operation of the system. (Courtesy of Compressor Systems, Inc.)

Skid-mounted packages can often be purchased for less than the cost of both the individual parts and installation. Also, skid-mounted compressor systems are well suited for portable applications in which the compressor is used for a time in one location then moved to another place.

Gas Sweetening Equipment

Gas that contains hydrogen sulfide is called sour gas. Hydrogen sulfide is a deadly posion and must be removed before natural gas may be used for residential consumption. Sweetening (the process of removing hydrogen sulfide) must be done on all sour gas. Sometimes this process is done in field production facilities, but it is usually delayed until the gas reaches a gasoline plant or refinery. Hydrogen sulfide and carbon dioxide are called acid gases because, when mixed with water, they form acidic solutions. Many pipelines require that acid gases be removed from natural gas before it is transported.

168 Introduction to Petroleum Production

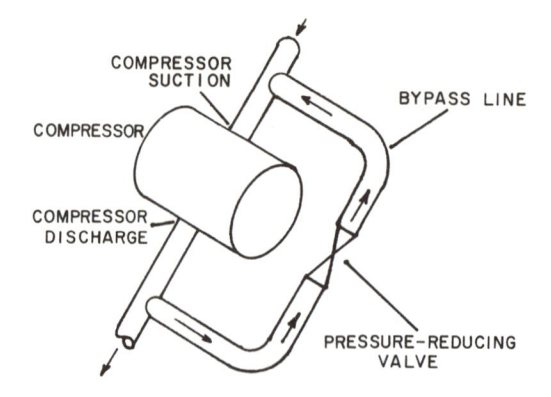

Figure 5-40. A by-pass line may be used to maintain enough suction to obtain maximum performance from a compressor.

Figure 5-41. Compressors can be furnished as skid-mounted systems complete with engine, cooler, piping, and control systems and requiring only inlet and outlet piping connections. (Courtesy of Compressor Systems, Inc.)

Sweetening equipment is often used to remove both hydrogen sulfide and carbon dioxide.

There are several processes which can be used to strip acid gas from natural gas. For field processing, however, the alkanolamine process is often used. This method uses a chemical of the alkanolamine family, such as amine, to strip the acid gases in a process shown in the simplified flow diagram of an amine sweetening unit in Figure 5–42. Sour gas enters the absorber tower and is forced into intimate contact with amine not contaminated with acid gas (lean amine) by a series of trays and baffles. The acid gas is absorbed by the amine, and sweet natural gas leaves at the top of the absorber. The foul amine then enters an amine stripping still which uses steam to heat the amine and distill the acid gas and water vapor from it. Acid gas and water vapor are cooled, and water vapor is condensed to its liquid state.

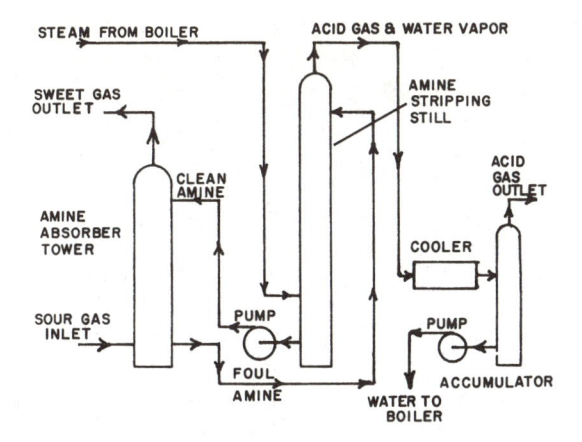

Figure 5-42. The alkanolamine sweetening process uses amine to remove acid gases (hydrogen sulfide and carbon dioxide) from natural gas.

Water is removed from the accumulator, while acid gas is vented to a flare where it is burned or is sent to a sulfur recovery unit which separates the elemental sulfur. Stripped amine is sent to a reboiler and reclaimer to complete the tasks of removing foreign substances from the amine, which is then sent back to the absorber as lean amine.

Figure 5-43 shows a skid-mounted sweetening unit. This photo shows all the equipment used in the process, some of which was not shown in the simplified flow diagram in Figure 5-42. Field gas sweetening units are usually designed for specific application, and the unit and vessel sizes vary with the volume of gas processed.

Sulfur Recovery Units

The effluent of a sweetening unit is a mixture of acid gas and water vapor. When small volumes are involved, the effluent is burned in a flare. The carbon dioxide and water, of course, do not burn; but combustion of hydrogen sulfide produces sulfur dioxide—a toxic gas only slightly less lethal than hydrogen sulfide. In many places the release of sulfur dioxide to the atmosphere is intolerable or restricted by government regulations.

Sulfur recovery units are process systems which remove elemental sulfur from the effluent of a sweetening process. Figure 5-44 is a simplified diagram of the process of removing sulfur. Acid gas is mixed with air and burned in a furnace to obtain a mixture of hydrogen sulfide and sulfur dioxide. This mixture is routed through a reactor where a catalyst causes a reaction of the gases to result in sulfur vapor. The sulfur is condensed to liquid and stored in a storage tank. Liquid sulfur can then be pumped to a sales point.

170 Introduction to Petroleum Production

Figure 5-43. Amine sweetening units are available as complete, skid-mounted assemblies that are trucked to a production site for immediate use. (Courtesy of Perry Gas Processors, Inc.)

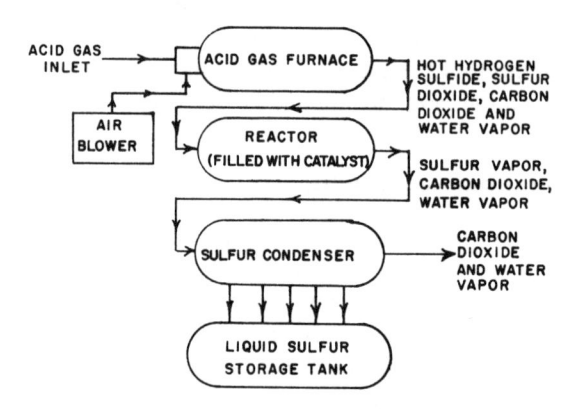

Figure 5-44. Sulfur is recovered from hydrogen sulfide using combustion and a chemical reaction with catalysts.

Sweetening and sulfur recovery processes are sometimes combined into a single operation. The equipment needed for both processes is then mounted in a single package as shown in Figure 5–45. In the last few years skid-mounted sulfur recovery units have been used in increasing numbers in both refining and field production facilities.

Gas Processing   171

Figure 5-45. Sulfur recovery systems can be provided to remove sulfur from all of the gas from a large gas field as it enters a gas processing facility.

Condensate Separation

When natural gas is produced, some of the lightest hydrocarbon liquids (pentane through heptane) are often present in appreciable quantities. These liquids are referred to as gas condensate and are valuable by-products of gas production because of their demand in petrochemical refining operations.

Condensate can be separated from natural gas with a two-phase separator when condensate is the only liquid present. If water is also present, liquid treating equipment (discussed in Chapter 6) may also be required.

Condensate is much more volatile than crude oil; that is, it can boil to vapor at lower temperature. When condensate is stored in tanks, the tanks must be completely sealed; otherwise, the vapors will escape and be lost. Tanks, which are discussed in Chapter 7, are normally equipped with vapor-tight covers called *thief hatches*. If thief hatches and all openings in a tank are tightly sealed, condensate may be stored without vapor loss.

Vapor Recovery From Storage Tanks

When petroleum liquids such as crude oil or condensate are stored in tanks, some hydrocarbons evaporate and stay in the gaseous state. If the tank is

172 Introduction to Petroleum Production

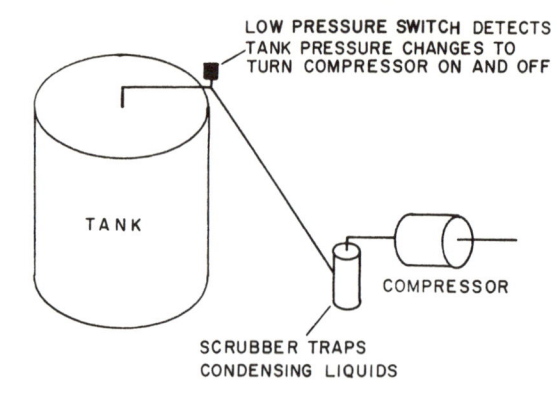

Figure 5-46. A vapor recovery unit senses internal tank pressure to turn on a compressor that removes natural gas from the tank.

completely sealed, evaporation continues, and pressure inside the tank increases slightly. At some pressure (a physical constant of the hydrocarbons), evaporation ceases. This pressure is only a few ounces per square inch.

Although the vapors released by evaporation are trapped inside the tank, unless some provision is made to pull these vapors out of the tank, some vapor will invariably be lost. A vapor recovery unit is a small, self-contained compressor designed to suck vapors out of tanks and pump them into the gas collection system.

Figure 5–46 is a diagram of a vapor recovery unit. The pressure switch is sensitive because it must respond to very small changes in pressure of a few ounces per square inch. The switch is adjusted to turn the compressor on when the tank pressure reaches a level slightly below the pressure at which evaporation ceases. The switch turns the compressor off when the pressure reaches some lower pressure just barely above atmospheric pressure.

Gas is sucked from tanks, through a suction scrubber, and into the rotary compressor. The compressor raises the pressure to no more than about 30 psig, unless multiple-stage compressors are used, and the gas is pumped into a low-pressure gas gathering system. Liquids trapped in the scrubber may be pumped into a low-pressure liquid system or back into the tanks. Figure 5–47 is a vapor recovery system with the necessary piping. The large suction pipe is sloped so that there are few restrictions and little pressure loss in it.

Gas Process Facilities

The gas processing facilities for a small lease could be configured as shown in Figure 5–48. Incoming fluids enter the facility at a header and enter a two-

Gas Processing 173

Figure 5-47. A vapor recovery system prevents the loss of hydrocarbon gases from tanks by removing the gas and injecting it into gas gathering systems.

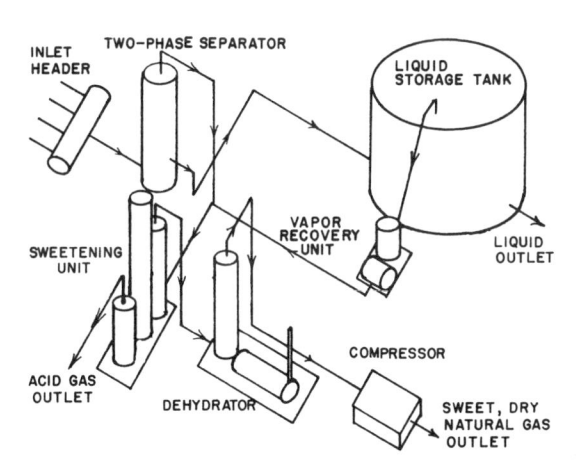

Figure 5-48. Field production gas processing facilities usually include two-phase separation, storage, vapor recovery and compression equipment, and may also include sweetening and dehydration equipment.

phase separator. Liquids are removed and stored in tanks. Gas is passed through a sweetening unit, then compressed and sent to a pipe line. This facility is representative of many gas process systems.

When both oil and water arrive at a gas processing facility or any other surface facility, the oil and water must be separated before the oil or condensate can be sent to refineries for further processing. The equipment used for separation of oil and water is usually considered a part of crude oil process facilities, but it can also be an integral part of gas treating systems. Chapter 6 discusses the equipment required to process petroleum liquids.

Chapter 6
Liquid Processing

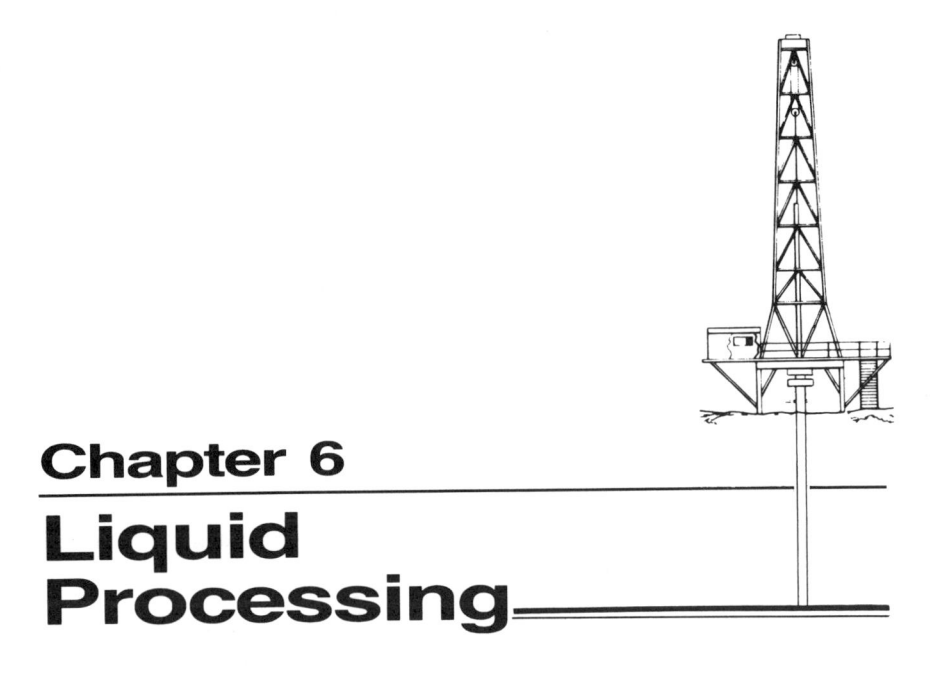

When oil and gas are produced with water, the fluid must be processed in a facility that separates the three fluids (phases). This separation is needed because most pipeline and refining systems to which oil is sent require that the oil contain less than 0.1–0.5% water. Also, the value of oil and gas depends on the amounts of light and heavy hydrocarbons, respectively. Petroleum liquids must be treated carefully to remove water and obtain the best possible blend of light and heavy hydrocarbons.

Oil-Water Emulsions

Oil and water are immiscible. When mixed, they exist as discrete droplets instead of a single, continuous mixture. As oil and water flow through the reservoir, wellbore, lift equipment, and gathering system, they form dispersions in which small droplets of one liquid are suspended in the other. An example of a dispersion is a vinegar and oil mixture used for salad dressing (See Volume 1, p. 126).

When emulsifying agents (some of the mild acids associated with production, iron sulfide, etc.) are present with oil and water, droplets form which have an internal phase completely surrounded by a skin composed of the other liquid and the emulsifying agents as shown in Figure 6–1. This is called

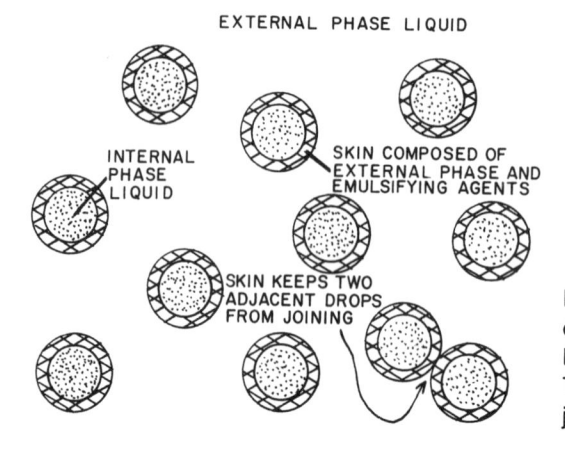

Figure 6-1. Emulsions are composed of droplets of one liquid suspended in another. Their skins prevent them from joining together.

an emulsion, and it is similar to the mixture of butter fat and water which compose homogenized milk.

Emulsions and dispersions are similar in that they are both composed of droplets of oil and water and, at least visibly, there seem to be no other chemicals in the mixture. They differ in droplet size; dispersions usually have larger droplets.

The tendency of a mixture of liquids to form emulsions or dispersions and the characteristics of these depend on liquid properties such as surface tension, density, viscosity, composition, and dissolved materials. The emulsions may be what are called "loose" mixtures (when the droplets are large and widely spaced), or "tight" mixtures (when the droplets are small and closely spaced). Most emulsions dealt with in petroleum production are called normal emulsions because water droplets are suspended in the continuous oil phase (a water-in-oil emulsion). Sometimes, however, oil droplets may be suspended in a continuous water phase. This is called a *reverse* or *oil-in-water emulsion*.

If an oil-water dispersion is allowed to stand long enough, the water droplets will coalesce (join together) into larger droplets which sink in the oil and eventually form a separate water layer below the oil. Some emulsions will also eventually separate into distinct layers when the emulsifying agent forms a weak skin, but most tight emulsions, like homogenized milk, can stand for quite some time without separating because the outer layer is too strong to allow coalescence.

Treating Emulsions

When oil and water form emulsions, some method must be used to separate the two liquids. Treating is the name given to the chemical and physical

ConocoPhillips Alok Jain

Date:

To:

ConocoPhillips

Alok Jain

Date:

To:

ConocoPhillips

Alok Jain

Date:

To:

ConocoPhillips

Alok Jain

Date:

To:

ConocoPhillips   Alok Jain

Date:
To:

Madras Pavilion   12 noon
Sunday   2 months old
Andy

1) AMSI is an incorporated entity & therefore the second clause will apply

2) AMSI

3)

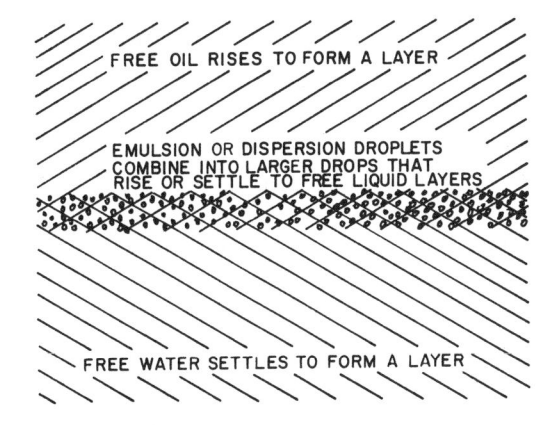

Figure 6-2. Gravity separation allows heavier water to settle below lighter oil while allowing droplets to coalesce and settle into layers.

processes used to separate liquids. There are three physical and chemical actions that can cause the separation of oil and water (a fourth action will be discussed in conjunction with electrostatic treaters).

Gravity Separation. Most dispersions and emulsions contain some free oil and free water that are not bound in emulsions. If the oil-water mixture is allowed to stand undisturbed, free oil rises to the top, free water settles to the bottom, and the dispersion will begin to separate as shown in Figure 6–2. As droplets approach, they collide and coalesce into larger droplets which settle or float into the oil and water layers.

Oil is lighter than water and would normally float in water. However, if the internal phase droplets are too small, they will continue to float in the external phase. Only when droplets collide and coalesce into larger droplets does the difference in specific gravity cause gravity separation.

The droplets of emulsions collide as they do in dispersions. If the skin is tough or if the surface tension of the external phase will not allow one droplet to join with another droplet, they can never coalesce into larger droplets. No matter how long they stand, most strong emulsions will not separate because of gravity.

Heating. The addition of heat to a dispersion speeds the separation process. In nature, all particles vibrate (including dispersion droplets), and increasing their temperature increases their speed. This increases the rate at which droplets collide and speeds the separation process.

The addition of heat to tight emulsions can cause oil-water separation. First, the increased temperature decreases the viscosity of oil and water and reduces the resistance to droplet movement. Second, added heat can weaken or eliminate altogether the skin surrounding internal phase droplets. Finally,

178   Introduction to Petroleum Production

increasing temperature increases the rate at which droplets collide. Thus, heat is one of the ways of breaking apart and separating emulsions.

Chemical Treating. Demulsifiers are chemicals that neutralize the effect of emulsifying agents. These chemicals breakdown the external phase of emulsion droplets and make the emulsion behave as a dispersion.

A number of different chemicals are blended to make demulsifiers. Since most emulsions differ slightly, the demulsifiers for each of these emulsions also differ somewhat. Demulsifiers are soluble in the external phase, so they come into intimate contact with all liquids.

Chemical Pumps. Demulsifiers are added to production streams in very small quantities of several quarts per thousand barrels—0.1% or less. A special pump, like the one in Figure 6–3, is used to pump demulsifier chemicals as well as other types of chemicals. This is a small, positive displacement pump. In many applications more than one chemical (demulsifiers, scale inhibitors, or corrosion inhibitors) is pumped at one location. Rather than use several chemical pumps, a single pump with several heads (Figure 6–4) is used.

Free-Water Knockout

Figure 6–5 is a cutaway diagram of a vessel called a free-water knockout. This vessel is intended to remove water that is not bound to oil in an emulsion. This water is called free water. The vessel is actually only a settling vessel because it allows liquid to stand quietly for a few minutes to allow water time to settle out of the dispersion by gravity separation.

Liquid entering the vessel separates into two distinct layers: one containing oil and emulsion, the other free water. Free water and emulsion are drained from the vessel into other vessels in the process facility by using float-operated dump valves similar to those used with two-phase separators. The float used to dump oil (the top float in Figure 6–6) is identical to those discussed in Chapter 5. The float used to remove water is called an *interface float* because it floats in water but sinks in oil. Interface floats are like air-filled floats, but they have weights inside to make them respond to the difference in specific gravities of water and oil. Each interface float must be weighted for a specific mixture of oil and water because there can be significant differences in specific gravities in each system. Figure 6–7 is a photo of an interface float which uses a counterbalance to allow adjustment from outside the vessel.

Figure 6-3. A small, positive-displacement pump is used to inject various chemicals into production streams.

Free-water knockouts are usually equipped with piping and pressure regulators, as shown in Figure 6–8, so that the vessel can be used to remove gas from a fluid stream as well as to separate oil and water. Although this is still a free-water knockout, it is often called a three-phase separator because it provides for separation of oil, gas, and water.

Fluids from a free-water knockout are usually piped to other vessels such as treaters or tanks. Rather than using pumps to move liquids, however, a difference in pressure between vessels is used to push liquids from one place to another. Thus, the internal pressure of a free-water knockout must be higher that that of the next vessel in sequence in the process facility. The internal pressure of a vessel is determined by the setting of its back-pressure regulator.

To separate as much gas from oil as possible, it is desirable to maintain the lowest possible pressure in the two- or three-phase separator. If this separator must dump liquid to another vessel, its pressure must be slightly higher to provide the driving pressure. Sometimes these two pressure objectives conflict, and care must be exercised to optimize pressures within a process facility.

Figure 6-4. Chemical pumps may be equipped with several heads to inject chemical into several streams or to inject several chemicals into the same stream.

Treaters

Oil-water emulsions are often processed with vessels called heater-treaters, emulsion treaters, or simply treaters. These are vessels which apply heat to emulsions to effect the separation of oil and water. In many cases demulsifiers are added to the liquid stream prior to entering the vessel. Also, the volume of treaters is large enough to allow the liquids to stand quietly for a time (the retention time is maximized). Thus, treaters use all three separation mechanisms: gravity, heat, and chemical separation.

Treaters are usually thought of as strictly liquid process vessels. When the temperature of oil is raised, solution gas is liberated, and some light hydro-

Figure 6-5. Fluids pass slowly through a free-water knockout and separate into oil, water, and gas layers.

carbon liquids are boiled into vapor. Thus, treaters also affect gas separation and must be equipped to handle this gas.

Vertical Treaters

Vertical treaters are the oldest type of treater and still predominate. There are slight variations in individual manufacturer's designs, but the general configuration of a vertical emulsion treater is shown in Figure 6–9.

Oil and water enter the treater in the top chamber through an inlet baffle. The inlet baffle causes agitation which aids in the separation of gas from incoming oil. The inlet liquid falls to the bottom of the vessel in a pipe called a *downcomer*. The downcomer is usually located inside the treater for maximum heat transfer, but it is sometimes located outside when corrosion or plugging problems require that it be easily accessible.

The liquids fall below the firetube into what is called the *free-water knockout section* or the *water bath*. Here, liquid stands long enough for water to settle to the bottom of the vessel while emulsion and oil float upward into the heater section. The heat from the firetube warms the emulsion, and water

182 Introduction to Petroleum Production

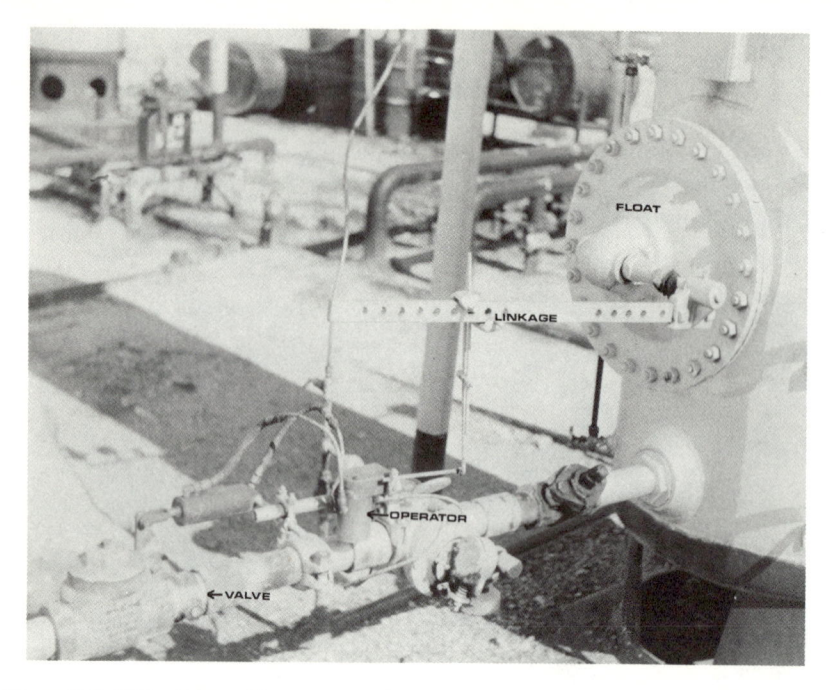

Figure 6-6. Oil is dumped from a vessel with a conventional float arrangement, while water must be removed with a different float system.

droplets coalesce and settle into the water bath while oil floats upward.

Oil is removed from the treater through a dump valve detailed in Figure 6–10. This valve is used to maintain a constant level in the treater. This level is located so that by the time oil floats to it, any water has had ample time to settle out. Water is siphoned out of the lower section through a *water leg* or *water siphon*. The water dump valve assures that water is removed at a fairly constant rate.

The firetube is like those used for production heaters and is designed to be in contact with emulsion or oil instead of very corrosive water. There is no need to heat the water, since it has already separated and would be a waste of heat energy. Because of the high temperature of more than 1000°F in the firetube, the firetube is subject to adverse conditions: heat stress caused by the heat, thermal expansion, and by oxidation inside and corrosion outside. Therefore, every possible effort should be made to keep the water level below the firetube to minimize corrosion.

In general, vertical treaters have a fairly short retention time of a few minutes, but they are able to apply a great deal of heat in a short time. Vertical treaters are also able to handle large gas volumes because the gas section may be as large as necessary.

Figure 6-7. A weighted float detects the interface between oil and water and allows water to be removed independently.

Liquid Level Controls. To maximize treating efficiency, the liquid levels in treaters can sometimes be critical. The oil level should be as high as possible to provide the longest possible retention time. The water bath must be high enough to provide adequate siphoning action but yet not reach the firetube.

The liquid dump valve shown in Figure 6–10 is a hydrostatically controlled valve instead of a float-controlled valve so that the valve and actuating mechanism are located at ground level for ease in adjustment maintenance. This valve type is used on most treaters, but it is also possible to use float-operated valves as shown in Figure 6–11. An unweighted float may be used for oil. Float valves have the disadvantage that they do not allow the level to be controlled over a wide range of several feet; however, they have the advantage of simplifying the piping—particularly the water leg piping.

Burner Controls. The burner used in the firetube is designed to release as much heat as possible when fuel gas is oxidized. The burner and the associated pilot are controlled by a fuel gas valve which operates as a thermostat. The fuel control assembly is shown in Figure 6–12.

The burner, firetube, and exhaust stack are shown in Figure 6–13. As fuel is blown into the firetube, air is pulled in and mixed with the fuel to form the

184 Introduction to Petroleum Production

Figure 6-8. A back-pressure regulator is used to maintain a constant pressure inside the vessel.

combustible mixture. When the firetube and exhaust stack have reached operating temperature, this high temperature causes a draft which pulls air into the tube. The entire combustion assembly is designed to work as a unit. As with natural gas heaters, treater firetubes are sized by the amount of heat they liberate internally. Two common sizes are 100,000 and 250,000 Btu/hr.

Occasionally, when a burner first ignites, gas which accumulates in the firetube can cause a flashback toward the ignition port. This can be dangerous for operating personnel and can damage instruments near the burner controls. Flame arrestors (Figure 6–14) are sometimes used to prevent a flashback.

Although a firetube may be operated with the burner running at all times, this is usually hard on the firetube because it is heat-stressed at all times. Most manufacturers recommend that a treater be operated so that the burner cycles on and off. That is, it comes on until the liquid reaches the desired temperature for a few minutes and then turns off for a time.

The temperature to which liquids are heated depends on the heat need to breakdown the emulsion. Some emulsions will separate at 80°F, while others require temperatures in the range of 140–180°F. As the emulsion is heated, light hydrocarbons are boiled out and leave the treater with the gas, decreas-

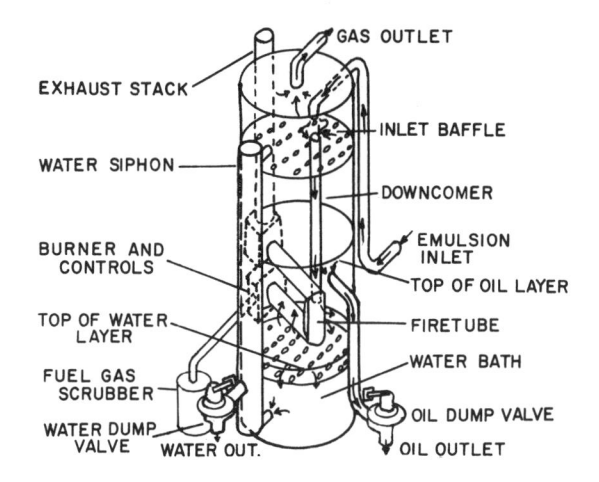

Figure 6-9. Vertical emulsion treaters use heat and the action of gravity to separate oil/water emulsions.

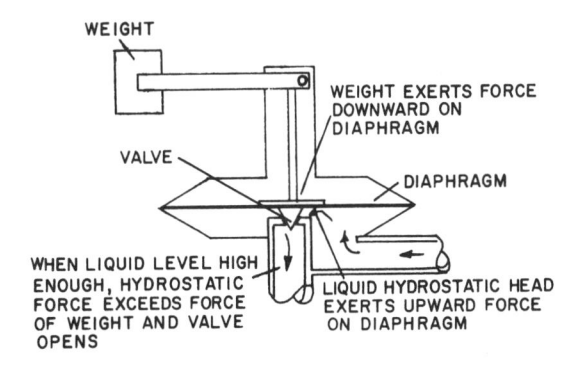

Figure 6-10. One type of oil dump valve senses liquid level by hydrostatic head and operates the valve.

ing the value of the crude oil. The lowest possible treating temperature should be used when separating oil and water. Many operators prefer increasing the amount of expensive demulsifiers so that lower temperatures can be used, and the value of the crude is maintained.

Gas Handling. The upper section of a vertical treater handles gas entering the treater. Gas enters this section in two ways. Incoming fluid strikes a baffle and lands on the upper plate. The downcomer is usually set above the bottom of the gas section so that liquid stands several inches deep. This allows gas to be liberated before falling to the water bath, and the volume of standing fluid absorbs slugs of liquid and maintains a steady downward flow. Gas liberated from emulsion by heat enters the gas section by the equalizer shown in Figure 6–15.

186 Introduction to Petroleum Production

Figure 6-11. A float-operated pneumatic dump valve may be used to remove oil or water from a treater.

The pressure inside a treater is determined by the back-pressure regulator because the pressure held in the gas section by the regulator is the same as the pressure throughout the vessel, neglecting the small hydrostatic effects. For liquid to be removed from the vessel, for the dump valves to operate properly, and for the water siphon to work, the pressure in the vessel must remain steady at a carefully selected pressure. Too much pressure can prevent fluid entry in the vessel, and too little pressure makes fluid leave the vessel very slowly if at all. Sometimes low pressure can cause the treater to flood—all the compartments fill completely and the siphon stops functioning.

When fluid enters the gas section rapidly or violently, a spray of liquid can be directed toward the gas outlet. A mist extractor prevents the liquid from

Figure 6-12. Thermostats control the operation of the pilot and main burner to maintain a constant temperature in a treater. (Courtesy of Engelman-General, Inc.)

leaving the treater via the gas line. Sometimes, a simple perforated baffle or screen around the gas exit is enough to prevent the spray from reaching the gas line.

Once fluid enters the water bath, every effort is made to make the liquids flow slowly and smoothly. Baffles are used sparingly if at all. One reason for keeping flow smooth is to allow maximum gravity separation. Another reason is that as emulsion is heated and gas liberated, oil-gas foams can develop. A foam can have undesirable results if it enters the oil dump line and valve, since the foam can make the valve "think" the oil section is empty when it is actually full to overflowing. Sometimes a wire sponge or mesh layer is placed in the upper part of the oil section in an effort to make the foam separate into gas and liquid.

Installation of Vertical Treaters. Vertical treaters are available in a range of sizes: 4–12 feet in diameter and 20–40 feet tall. The treater is usually mounted on a skirt as pictured in Figure 6–16. A vertical treater is mounted on a flat concrete base as shown in the same figure. For some smaller treaters

188 Introduction to Petroleum Production

Figure 6-13. The main burner, pilot, firetube, and exhaust stack are readily accessible but are covered to protect personnel.

Figure 6-14. Flame arrestors are used on most treaters to keep flames from flashing back at an operator as the burners ignite.

installed in areas of low wind velocities, the treater is simply set upon the base, and the weight of the vessel and its contents hold it in place. Larger vessels or those used in areas of high wind are anchored to the concrete base, which itself is anchored. When necessary, guy wires are installed.

Fluids can sometimes flow in slugs into a treater. Since vertical treaters are already somewhat top-heavy, slugging flow can cause the treater to rock. However, if it is installed on a firm footing and adequately anchored, there is no danger of the vessel falling.

Horizontal Treaters

A horizontal treater (Figure 6–17) is used to separate an emulsion into its component parts by adding heat to the liquid. Horizontal treaters usually have a longer retention time because they are usually larger than their vertical

190 Introduction to Petroleum Production

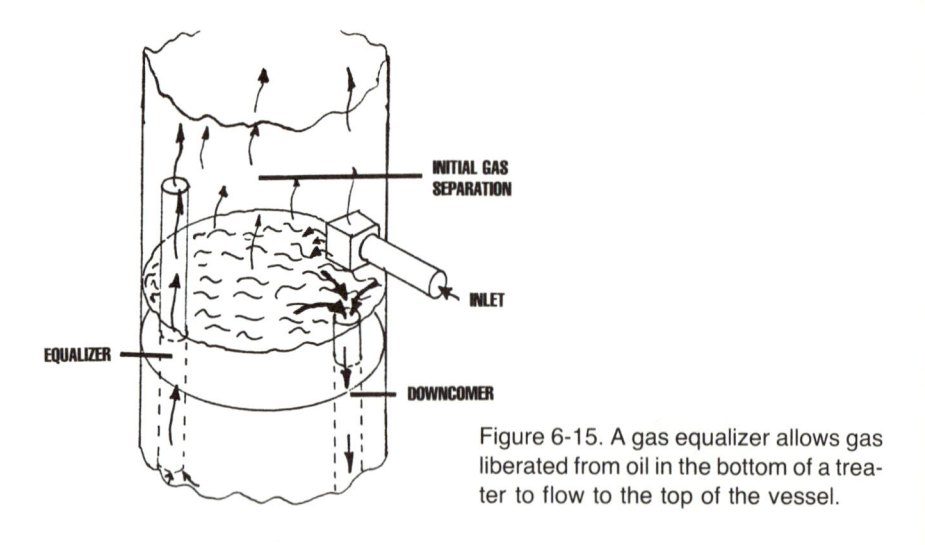

Figure 6-15. A gas equalizer allows gas liberated from oil in the bottom of a treater to flow to the top of the vessel.

Figure 6-16. Treaters are mounted on skirts to facilitate standing without violating the spherical vessel shape needed for high-pressure construction. (Courtesy of Midessa Equipment Company.)

Figure 6-17. Horizontal treaters are used when long retention times are required and when space is at a premium.

counterparts. Large horizontal treaters may be 10–20 feet in diameter and 30–50 feet in length.

Fluid enters the vessel through a set of baffles where free water is allowed to settle immediately and flow into the bottom rear of the treater. Emulsion spills into the area adjacent to the firetube and is heated. Oil and the remaining emulsion flow over the top of this heated area and into the rear of the vessel. By the time oil and emulsion reach the rear of the treater, they are still quite warm (90–140°F depending on the thermostat setting). Liquids are retained in the rear of the treater long enough for the remaining emulsion to breakdown and separate into oil and water.

Gas leaves the treater through a mist extractor and back-pressure regulator. Oil is removed from the vessel through a dump valve which may be hydrostatically or float-operated. The water dump valve is usually controlled by a weighted interface float.

192 Introduction to Petroleum Production

Figure 6-18. Horizontal treaters may be equipped with gas domes to increase their gas processing capacity. (Courtesy of Midessa Equipment Company.)

Some horizontal treaters are equipped with a gas dome (Figure 6–18). This allows additional space for inlet baffles and a mist extractor. More important, with the inlet baffles and gas outlet facilities outside the body of the treater, liquid levels in the vessel can be higher to provide more retention time and more space for gas flow. This higher-efficiency treating increases the amount of gas that can be processed.

The rear section of a horizontal treater has no heat added to it, but the liquid is expected to retain its warmth as long as possible. For this reason, most horizontal treaters are heavily insulated (most vertical treaters are not insulated) to prevent heat loss through the vessel walls. The insulation first used was a polyurethane foam, but this has been replaced by a glass wool in most applications because polyurethane gives off a poisonous gas if it is burned. Several inches of insulation are placed around the treater body and then covered with a galvanized steel or aluminum skin as shown in Figure 6–19.

Horizontal treaters are usually used when long retention times are required and gas volumes are low (the gas capacity is increased by the addition of a gas dome). The cost of horizontal treaters is high compared to vertical treaters. The mechanical arrangement of horizontal treaters makes them well suited for applications where space is critical, such as an offshore facilities, or when they must be enclosed in a building in populated areas in some states.

Liquid Processing 193

Figure 6-19. Treaters are usually provided with wool insulation covered by a layer of aluminum to keep from losing heat to the atmosphere. (Courtesy of Midessa Equipment Company.)

Electrostatic Separation

The application of electrostatic fields is another frequently used method of separating emulsions. If an emulsion is forced to flow between two plates which have a high voltage across them, the droplets become polarized and their shape distorted as shown in Figure 6-20. Electrostatic dehydration uses two grids and impresses high voltage (60 cycles per second) electric current on them. When many emulsion droplets are between these grids, the droplets are distorted and polarized, return to their original shape, and are again distorted but polarized in the opposite direction 60 times per second as shown in Figure 6-21.

When droplets are polarized (electrically charged) they attract each other. This attraction causes the droplets to collide and coalesce as shown in Figure 6-22. As droplets combine, they become large enough to settle to the oil and water layers by gravity action.

Electrostatic dehydration can be used only when oil is the continuous phase of the emulsion. Because oil is an excellent insulator, little current flows between the grids when the voltage between them is high (20,000 volts). However, if water (a good conductor of electricity) is the continuous phase between the grids, the grids short out, and the process stops.

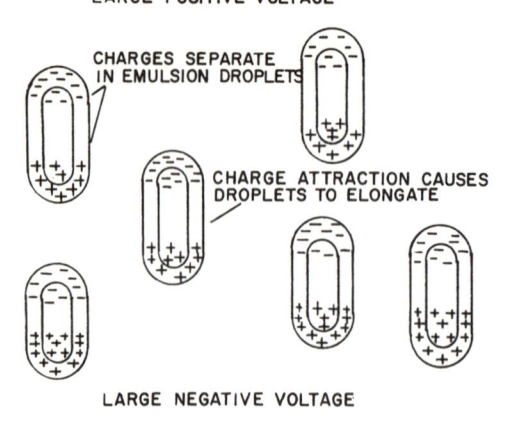

Figure 6-20. Under the influence of electrostatic fields, emulsion droplets become polarized and distorted in shape.

Electrostatic Treaters

Figure 6–23 shows a treater in which electrostatic separation is used as one of the methods of breaking down an emulsion. Liquid enters the vessel and flows around the firetube where it is heated. Water flows to the bottom rear of the vessel, while emulsion and free oil flow into the area between the grids in the rear section.

The high-voltage grid is connected to the secondary of a transformer, and its voltage with respect to ground is 10,000–25,000 volts. Since the body of the treater is grounded, there are two areas where a high voltage exists. One is from the high-voltage grid to the top of the water layer, which is electrically connected to the treater's ground potential by its own conductivity. Emulsion in the area between the water layer and the ground grid is subjected to high-strength electrostatic fields.

In an electrostatic treater emulsions are treated by all four separation mechanisms, but the electrostatic field is the dominant mechanism. The addition of demulsifiers weakens the skin surrounding emulsion droplets, while addition of heat reduces the viscosity, making it easier for the droplets to collide with each other.

The electric power consumed by an electrostatic dehydrator is so small as to be negligible because almost no current flows through the grids. However, the grids' effect means little heat is needed (treating temperatures less than 80°F are not uncommon), and the required demulsifier rate is also very low.

One of the most important advantages of electrostatic treaters is the low treating temperature. When the fluid temperature is 80–100°F, there is almost no evaporation of light hydrocarbons. Thus, the value of the crude oil is enhanced.

Liquid Processing   195

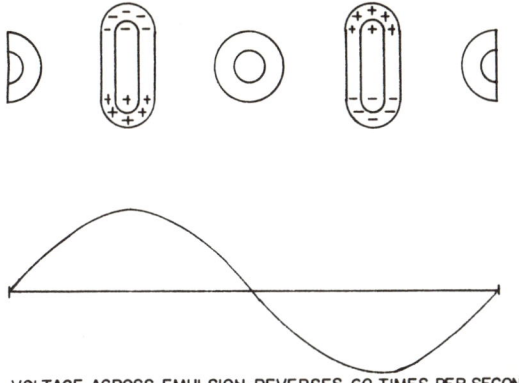

VOLTAGE ACROSS EMULSION-REVERSES 60 TIMES PER SECOND

Figure 6-21. Under the influence of an alternating voltage, emulsion droplets alternately polarize in one direction and distort, return to normal, polarize in the opposite direction and distort, and return to normal again.

OPPOSITELY CHARGED DROPLETS ATTRACT EACH OTHER   WEAKENED SKINS ALLOW DROPLETS TO JOIN   DROPLETS COALESCE

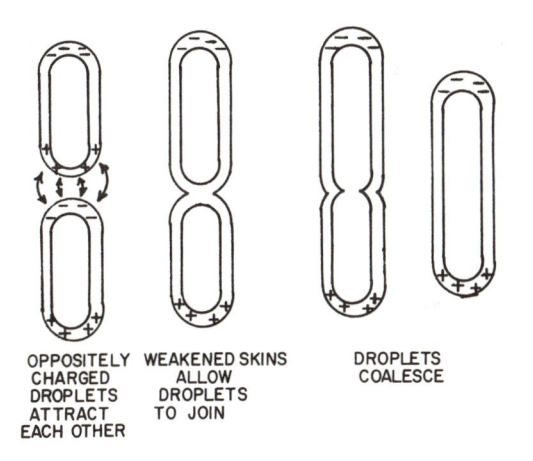

Figure 6-22. Polarity caused by eletrostatic fields makes droplets attract each other, collide, and coalesce into larger droplets.

An electrostatic treater is pictured in Figure 6–24. A step-up transformer on top of the treater supplies the high voltage to the grids through high-voltage entry bushings. The body of the treater is heavily insulated to hold heat inside it. During warm summer months, many operators never turn on the burners because the ambient temperature is enough to treat emulsions.

An electrostatic treater utilizes the same controls as any other treater. A back-pressure regulator is required to maintain pressure inside the vessel.

196 Introduction to Petroleum Production

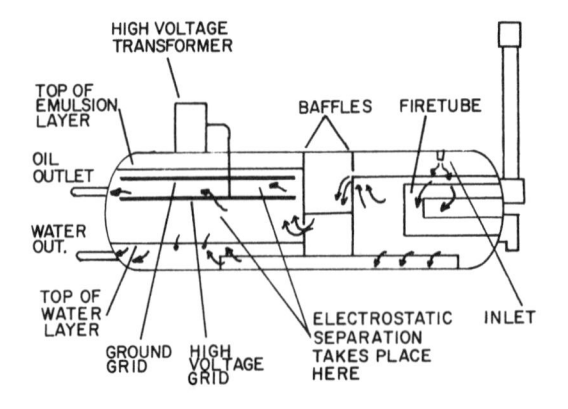

Figure 6-23. An electrostatic treater uses electrostatic separation with heat to treat emulsions.

Figure 6-24. An electrostatic treater has a firetube like conventional treaters and a high-voltage transformer powering internal grids for electrical treating.

The dump valves may be operated by floats as shown, or hydrostatic dump valves as described for vertical treaters. Gas domes are used to increase the gas capacity.

Safety Precautions With Treaters

The firetube of all treaters is exposed to conditions that can cause failure. The pressure inside the firetube is low—near atmospheric pressure—while the pressure outside is the internal pressure of the vessel (10–100 psig). Any leak causes oil and emulsion to flow into the firetube. Crude oil is not particuarly volatile at ambient temperature, but when oil enters the hot firetube, it will ignite almost immediately. Firetube fires are the most serious failure that can happen in a treater. If the fire is not extinguished quickly, a very dangerous fire can develop.

Many treaters are insulated with polyurethane compounds. When heated or burned, this insulation gives off very poisonous gases. When a treater begins to burn, these gases present a threat to personnel fighting the fire. Most recently constructed treaters use glass wool insulation to avoid this problem, but precautions should be taken with every treater fire to avoid the danger of entering a lethal atmosphere.

Treater firetubes can be protected to minimize the possibility of leaks. One way is to coat the outside of the firetube with specially formulated, high-temperature resins and epoxy. This coating keeps corrosive liquids from contacting the firetube and causing corrosion failures in the metal.

When the burner ignites, the firetube rapidly heats and expands, but it contracts when the burner is turned off. The continued expansion and contraction can cause fatigue failures in the firetube. Firetubes can be designed with expansion joints and firm bracing to minimize movement and prevent fatigue failures.

A precaution often taken with any vessel using firetubes is the installation of flame arrestors (Figure 6–12). These are used for both equipment and personnel protection. Some burner assemblies have arrestors built into the assembly, while others require the arrestor as an addition.

The gas used for fuel in treaters can pose a danger before and after combustion. Fuel gas is usually taken from the produced gas from a lease. If produced gas is sour, it should be sweetened before using it as fuel. The hydrogen sulfide is a lethal poison and could pose a danger to operating personnel if the burner control valves do not shutoff fuel completely. When hydrogen sulfide is burned, it produces sulfur dioxide—a somewhat less lethal but still poisonous gas. When produced gas is not sweetened, it is often necessary to secure sweet fuel gas from an outside source.

API Gravity

After crude oil passes through a treater, it should have much less than 1% water, but still contain a mixture of hydrocarbon liquids. Specific gravity is one way of expressing the composition of the treated oil. Another way of expressing this weight is the API gravity of oil. API gravity expressed in degrees can be obtained from the expression: API Gravity = 141.5/Specific Gravity − 131.5

198   Introduction to Petroleum Production

For example, the API gravity of pure water (specific gravity of 1.0) is 10 degrees.

As the proportion of light hydrocarbons in crude oil increases, the specific gravity of the mixture decreases, and the API gravity increases. The price of crude oil is determined by its API gravity, and the price increases with API gravity. High-gravity crude oil is more valuable as refinery feedstock than low-gravity crude because more light products such as gasoline and naphtha can be made from high-gravity crude. Thus, it is to the producer's advantage to maintain the highest possible API gravity.

Treating Optimization

The processes required to separate oil and water are costly. Fuel gas is costly whether it comes from the lease itself (because it cannot be sold as produced natural gas) or from an external source (because it must be purchased). Demulsifiers are also quite expensive as is electricity when electrostatic treating is used. In addition, labor costs to operate treating facilities have skyrocketed in the last few years. In every treating application the operating methods must be optimized to maintain the lowest treating cost while keeping the API gravity as high as possible.

Some emulsions are very difficult to breakdown. One way of attacking these mixtures is to use high-temperature treating—bringing the emulsion temperature to 120-140°F. In general, this temperature will separate almost any emulsion without requiring significant volumes of costly demulsifiers. However, this high temperature requires the use of large volumes of fuel gas. High-temperature treating boils off light hydrocarbons and decreases the API gravity. High-temperature treating also places a great deal of thermal stress on the firetube and eventually causes failures, thereby increasing maintenance costs. However, this method of treating is often used because the initial treater cost is usually fairly low.

Tight emulsions can be treated at much lower temperature if large volumes of demulsifier are used. Of course, the fuel cost and maintenance costs are reduced when low-temperature separation is used, but the chemical cost increases dramatically. Because the temperature is lower, few hydrocarbons boil out of the oil, and its gravity is usually high.

If very large treaters are used, low temperatures and moderate chemical rates can be used because the retention time in the vessel is increased. The long-term operating costs are usually lower than for either of the previous alternatives, but the initial cost of the vessels is high.

Electrostatic treating also offers an alternative to treating costs. Electrostatic treaters usually operate at low temperatures and require little assistance from demulsifiers. The power cost is appreciable but usually lower than fuel or chemical costs. The initial investment in an electrostatic treater is greater than for comparably sized conventional treaters.

There is no "best" way to treat emulsions. A separation system must be carefully designed for each production application. The initial selection of equipment must be done with future operating costs in mind. Treater insulation and even the vessel color must be considered (light-colored vessels absorb less radiant energy from the sun than dark-colored vessels) to try to maintain liquid temperature for least-fuel consumption. Burner design and arrangement and firetube configuration should be studied to obtain the highest energy conversion efficiency and minimize future firetube failures. Many operators consider future operating costs and treating efficiency to be far more important than the initial cost of the treating system.

Other Treating Methods

Treaters are not the only equipment that can be used for separating oil and water. Because they are compact, self-contained, and easy to install and operate, treaters are the most common method of removing water from oil. However, open-pit treating and gun-barrel tanks are still used occasionally.

Open-Pit Treating

This was the original method of removing water from oil for many years before treaters were developed. Emulsion is dumped into an open pit—usually a plastic-lined, earthen or concrete pit. These pits are large (several thousand square feet in surface area), shallow (10 feet or less) holes in the ground.

The treating mechanisms involved in open-pit treating are heating and long retention time. The sun's radiant energy is absorbed by the liquid in the pit, and the liquid warms to at least the ambient atmospheric temperature. The warm liquid stands quietly in the pit for several hours to several days. The emulsions breakdown and separate into oil and water layers, and these liquids may then be pumped out of the pit.

Open pits are the least satisfactory of all treating methods and are seldom used now. Hydrocarbon evaporation rates are high, and a large proportion of the light and medium weight hydrocarbons are irrecoverably lost. Exposure to direct sunlight and air can create a skin on emulsion droplets, and any emulsion that does not separate can require extensive treating. Open pits are also a fire hazard because oil is directly exposed to and mixed with air, and at the surface of the liquid an explosive mixture can exist. Open pits have been outlawed in many states because of atmospheric, surface, and subsurface pollution. It is almost impossible to keep liquids from seeping into the earth from even the best lined pit.

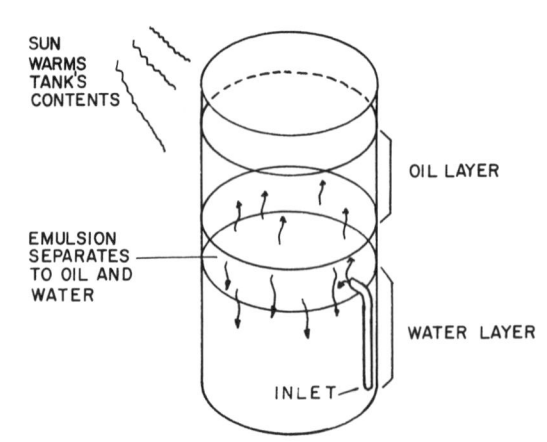

Figure 6-25. A gun-barrel tank uses the force of gravity and some radiant heat to separate emulsions.

Figure 6-26. Gun-barrel tanks are tall tanks used to provide long retention time in treating operations.

Gun-Barrel Treating

Another method of treating involves the use of tanks. These tanks are tall, slender tanks or conventionally sized tanks. Liquid is pumped into a tank and allowed to stand quietly as shown in Figure 6–25. The tank is warmed by the sun, and this warmth and long retention time causes emulsion to separate into layers of oil and water.

Gun-barrel tanks are often used in modern treating facilities to allow some separation before emulsions reach treaters as shown in Figure 6–26. Warm water that has already been separated from emulsions is sent to a gun-barrel tank where it forms a water bath. Incoming produced fluid is forced to flow through this warm water bath and stand for a short period in the tank. By the time the emulsion reaches the treater, a great deal of water has already separated from the oil. This reduces the requirements of the treating system and improves the efficiency of the whole facility.

Oil and Water Handling

After an emulsion has been treated, the oil and water must be sent somewhere. In some very isolated cases a treater dumps directly to the pipeline where oil is sold. In the vast majority of cases the oil and water must be dumped to storage tanks and then pumped to whatever destination is desired. Chapter 7 discusses storage and disposition of fluids in petroleum production facilities.

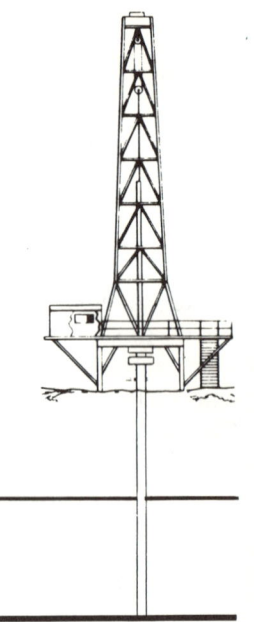

Chapter 7
Liquid Storage

Once incoming fluid has been separated into oil, water, and gas, these fluids must be either stored or transported. Gas is seldom stored on the lease because gas storage facilities are usually far too expensive to be justified on a small-scale basis. However, oil and water are usually stored for some time in tanks.

Liquid Storage Tanks

There are several types of tanks used for liquid storage. Originally, redwood tanks were used for storage because they were the most reliable type of tank at the time. Wooden tanks are no longer used and have been replaced by steel tanks.

Steel Tanks

Steel storage tanks are available in two forms: (1) small tanks (500 barrels or less), and (2) large tanks (1000 barrels or more). Small tanks are usually of welded construction as shown in Figure 7–1. They are easier to build in welded form because the sections of the tanks can be preformed and assembled in a fabrication yard. Larger tanks of 1000 barrels or more are usually of bolted construction. Large sections of the tank are brought to the site

Figure 7-1. Welded tanks are common for sizes up to 1000 barrels.

and assembled on location. The sections are simply bolted together as shown in Figure 7–2.

Solid sediment produced with oil and water usually accumulates in tanks because this is the first place where fluid is stopped long enough for the solids to settle. The shape of the tank bottom is important in the movement of solids. Flat-bottom tanks (Figure 7–3) were the first tank shapes developed. When solids present no problems in a facility, flat bottom tanks are satisfactory for use. Cone-bottom tanks (Figure 7–4) are used most often for permanent storage facilities because they are better suited for handling solids. As solids settle from liquids, they drift to the center of the bottom of the tank. Since the solids are always located in the same place, it is easier for them to be recovered or at least kept suspended in the liquid. Periodically, solids must be removed from tanks. Solids may be pumped out with some liquids, or people may enter the tank and scoop up the solids.

On many production leases, storage requirements do not remain fixed for long periods of time. Sometimes a water tank is required for several months but then not required again for several years. Many production facilities are built using temporary or portable tanks. That is, tanks are used for a time and

204 Introduction to Petroleum Production

Figure 7-2. Bolted tanks are available for small volumes but are usually used for large-volume applications.

SOLIDS ACCUMULATE IN FLAT-BOTTOM TANKS

LITTLE SITE PREPARATION REQUIRED

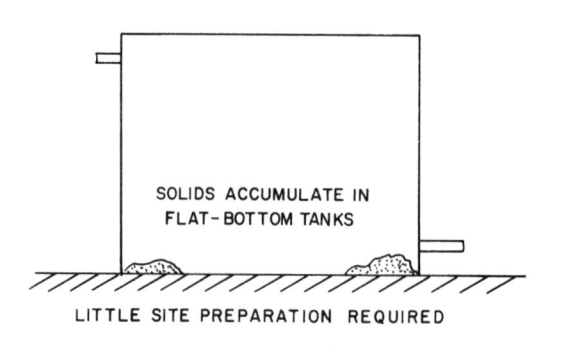

Figure 7-3. Flat-bottom tanks require little site preparation but have a tendency to collect solids.

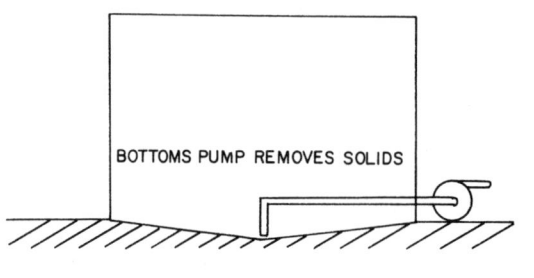

BOTTOMS PUMP REMOVES SOLIDS

TANK BASE MUST MATCH CONE SHAPE OF TANK BOTTOM

Figure 7-4. Cone-bottom tanks require site preparation but have little tendency to collect solids.

Liquid Storage 205

Figure 7-5. Tanks are usually mounted on raised tank pads for additional support and to facilitate cleaning and maintenance.

then moved to another site. Flat bottom tanks are usually used for temporary applications because they can be installed with a minimum of site preparation.

Site Preparation

Storage tanks are very heavy when filled with liquid (a 500-barrel tank filled with water weighs more than 80 tons). The earth on which a tank sits must be carefully prepared to be able to withstand this load without shifting.

Most permanent storage facilities are built on a pad or raised surface as shown in Figure 7-5. It is easier to form the tank base on a raised surface than at ground level. The tank pad must be formed of firm material that has been leveled and thoroughly compacted. Then as the weight of a tank bears down on the pad, the pad is not likely to shift.

The tank pad must correspond to the shape of the tank bottom exactly. If the base under a tank is uneven, the weight of liquid will exert uneven forces on the tank bottom and sides. The sides of storage tanks are not designed to withstand vertical forces, only lateral forces. Therefore, the tank base must be shaped to correspond to the bottom of the tank.

In populated areas and critical storage areas tank dikes are often required by law but should be used as a part of sound operating practice. A tank dike, shown in Figure 7-6, is a raised earthen structure with a volume at least as large as the tank's. The dike's purpose is to contain the contents of a tank in the event of a massive leak and keep the liquid from running into populated

206   Introduction to Petroleum Production

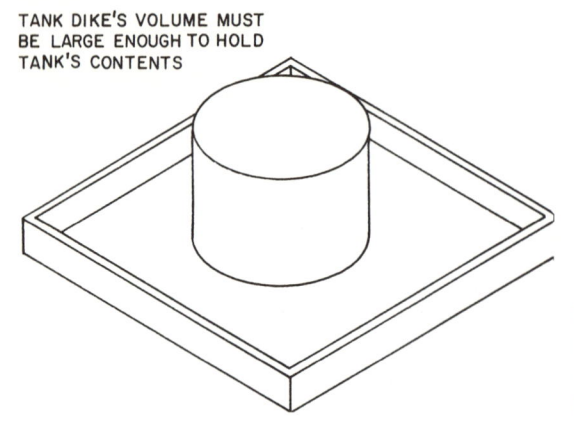

TANK DIKE'S VOLUME MUST
BE LARGE ENOUGH TO HOLD
TANK'S CONTENTS

Figure 7-6. A tank dike is used to hold the contents of a tank in case of a leak or a fire.

areas or polluting the surrounding ground. These dikes are particularly effective if a tank ruptures and the contents ignite. The burning liquid is restricted to a small area where it can be extinguished easily.

The tank dike is sometimes constructed with fire fighting in mind. A pipeline equipped with spray nozzles is run around the dike (Figure 7–7) and connected to a water or foam supply. This design provides for sprays on the tanks to extinguish a fire or cool the tank to prevent fire.

Tank dikes may be made of earth shaped with earthworking machinery. This earth is usually covered with asphalt to prevent erosion. The dike may also be constructed of concrete for sturdy and permanent application.

Ancillary Equipment

In addition to the body of the tank which actually contains the liquid, there are several pieces of equipment associated with storage tanks that improve or augment the operation of the tank system.

Sumps. Sumps (Figure 7–8) are sometimes used with tanks to allow a place for solids to accumulate and to allow a place to pump liquids. The sump is simply a large-diameter cylinder (12–36 inches) at the bottom of the tank. Liquids are pumped out of the tank from a suction line in the sump.

Risers. Risers are sometimes used to force liquids to enter or exit a tank from a level different from the level at which the liquid enters the tank. Figure 7–9 shows the use of a riser to let liquid entering the tank at ground level actually flow into the tank at a higher level.

Figure 7-7. Fire spray nozzles are sometimes installed in critical areas around the dikes of tanks should fire-fighting techniques be needed.

Siphons. Sometimes siphon lines (Figure 7–10) are used to remove liquid from the bottom of a tank. The siphon line utilizes the same physical principles as common applications of siphoning.

Spreaders. When liquid enters a tank (like a gun-barrel tank), the liquid is expected to enter as a spray instead of a liquid stream. A spreader, such as the one in Figure 7–11, makes liquid enter over a wide area. Spreaders are used to make incoming liquids mix with stored liquid or aerate incoming liquids.

Roll Lines. Even though solids do accumulate in the bottom of tanks, it is sometimes undesirable to allow them to do so. A roll line (Figure 7–12) is a thin pipe which is routed around the inside circumference of a tank. The pipe is perforated so when gas is injected into the tank, it agitates the tank's contents.

Agitators. An agitator is a propeller or fan inside the tank. The propeller is rotated by an external electric motor. Agitators are used to move liquids within the tank and agitate solids and liquids.

Thief Hatch. Access to the tank's interior is necessary to determine its contents. Tanks are equipped with thief hatches which allow access, but which

208 Introduction to Petroleum Production

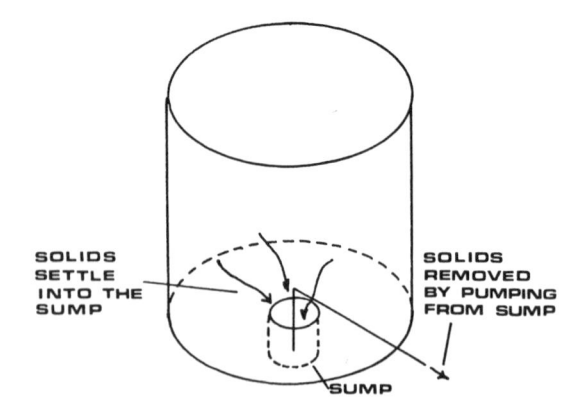

Figure 7-8. Solids that accumulate in a tank settle into a sump from which they may be removed by pumping.

Figure 7-9. Internal or external risers are used to make liquid enter a tank at a high level.

Liquid Storage   **209**

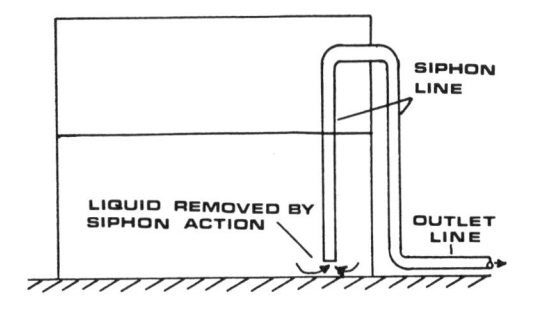

Figure 7-10. Liquid is sometimes removed from a tank by simple siphoning.

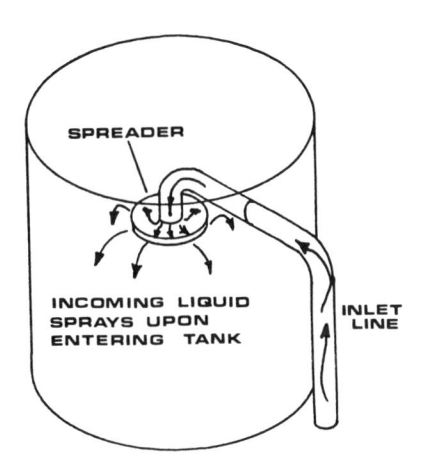

Figure 7-11. Inlet spreaders make liquid enter a tank in a widely distributed spray rather than in a single stream.

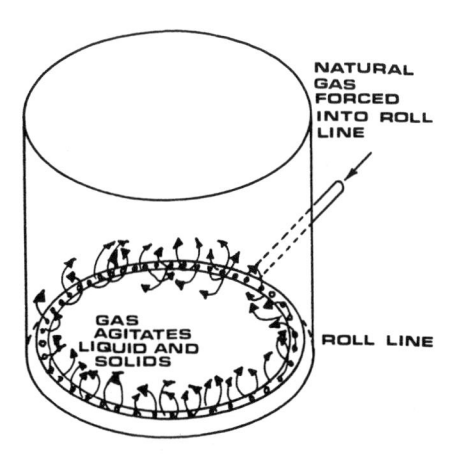

Figure 7-12. Gas injected through roll lines agitates solids so they can be removed with pumps.

seal the tank when closed. A thief hatch (Figure 7–13) is a pressure-actuated device which can be opened but which also acts as a seal and an over-pressure protection device. Because tanks have large areas exposed to internal pressures, the internal pressure must be kept lower than a few ounces per square inch to keep from rupturing the thin metal walls and deck.

Pressure Relief Valves. Although thief hatches serve as high-pressure relief valves, their adjustment is somewhat unpredictable. Relief valves are used when the venting pressure is critical. One application of relief valves is when vapor recovery units are used. The internal pressure is maintained as high as possible when vapor recovery units are used to provide maximum suction pressure, but it is important that a relief valve be used to keep the pressure from becoming excessive.

Walk and Stair. Occasionally, it is necessary to reach the top of tanks to adjust valves or sample the contents. Walkways and stairs are necessary to reach and safely work at the top of the tanks. A typical walk and stair system is pictured in Figure 7–14. The Occupational Safety and Health Administration (OSHA) has a number of regulations covering safety in industrial environments. OSHA specifications cover walk and stair and require safe and sturdy structures.

Corrosion Protection

Tanks are often required to store corrosive liquids (such as water) with dissolved gases (such as oxygen, carbon dioxide, or hydrogen sulfide). Since tanks are erected for long periods of service, every effort should be made to prevent corrosive attacks on the metal.

Tank Coating. Most metal tanks are internally coated with plastic or epoxy compounds that are intended to adhere permanently to the interior surfaces. Tanks are sand blasted thoroughly to remove any oxide coating on the metal. The coating material is then sprayed onto the interior surfaces and allowed to air cure for several days. After curing, the coating is checked for holidays (bubbles, breaks, or other imperfections in the coating layer). Coating may be applied before a small tank is erected, but it is usually applied after the tank has been erected in place. Lifting, transporting, and setting a tank in place can crack a coating that has been previously applied.

Fiberglass Tanks

Preformed fiberglass tanks are available for many small-volume applications of up to about 250 barrels. Open-top water tanks are inexpensive and

Liquid Storage 211

Figure 7-13. A thief hatch allows access to the inside of a tank but acts as a pressure relief when closed.

easy to install on a prepared earth surface. Larger, sealed tanks are erected with slightly more effort. Fiberglass tanks are usually used for water storage because no provisions are needed for corrosion protection.

Open-Pit Storage

Water is often stored in open pits when the water is to be disposed of shortly. Open pits are used because some of the water evaporates naturally. Water pits may be earthen, but they must be lined with plastic or asphalt to keep the water from seeping down and contaminating shallow fresh-water aquifers. Pits may also be constructed of concrete, but even the concrete must be sealed with plastic to prevent water from percolating through the bottom of the pit.

Level and Sample Measurement

The height at which liquid stands is a measurement of the volume of liquid in the tank. When the depth is being used to calculate the precise volume for sale of oil, a gauge line (Figure 7–15) is used (as shown in Figure 7–16). The

212 Introduction to Petroleum Production

Figure 7-14. Walk and stair provides a safe and convenient method for operators to reach equipment mounted on top of tanks.

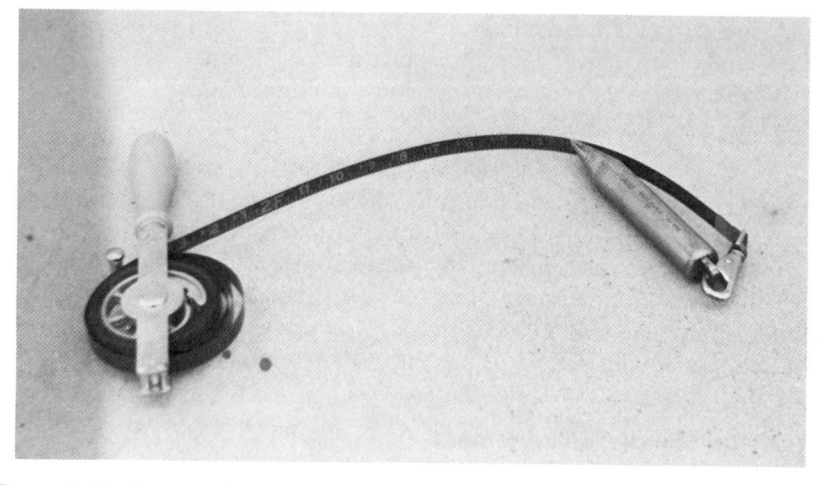

Figure 7-15. A gauge line used to measure liquid depth in a tank is similar to a flexible tape measure.

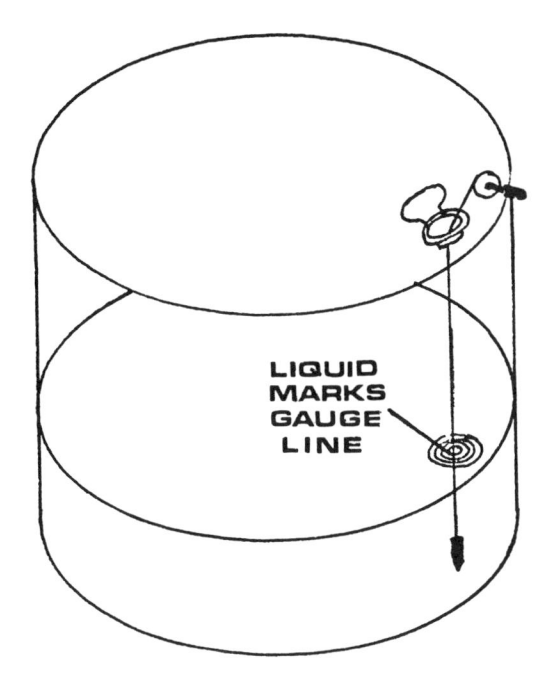

Figure 7-16. When a gauge line is lowered into a tank, the liquid leaves a mark on the line that can be read when the line is reeled back to the top.

weight is lowered into the tank until the weight hits the tank bottom. Oil or water coats the gauge up to where the liquid level is.

When only a quick check is needed to determine the tank level without measuring to the nearest fraction of an inch, two methods are available. A float mechanism, like the one in Figure 7–17, is an inexpensive method of showing the level in a tank. Another, more sophisticated and accurate method is to use a head gauge as shown in Figure 7–19. The hydrostatic head exerted against the diaphragm actuates a pointer which moves along a calibrated scale. An operator may read the level in a tank to a remarkably accurate degree without climbing the tank and running a gauge line.

Tank Sampling. On many small leases, crude oil is produced into a tank until it is full. Production is then routed to another tank while the original is measured and sold. Liquid inside the tank is sampled to determine the volume, the amount of water still contained in the oil, and the gravity of the oil. When these measurements are complete, the tank valves and thief hatch are closed and sealed. An embossed aluminum strip or wire is attached to the

214 Introduction to Petroleum Production

Figure 7-17. An external float may be used to provide a visual indication of the depth of liquid in a tank.

hatch which, if unbroken, indicates the equipment has not been tampered with.

Liquid in a tank is sampled with a device called a *tank thief* (Figure 7–18). A thief is a sealed sample container with a valve which may be opened with a rope. An empty thief is lowered into a tank and its valve opened to trap a sample of the liquid. Samples which represent liquid at several levels in the tank are usually taken.

When a sample is recovered from a tank, it must be analyzed to determine its composition. The gravity of a sample is determined with a hydrometer (Figure 7–19), a weighted float. The hydrometer's submergence depth depends on the specific gravity of the liquid (Figure 7–20).

The composition of a sample is measured by using a centrifuge. A centrifuge, such as the one shown in Figure 7–21, uses centrifugal force to separate a small sample of emulsion into its constituent parts. A portion of the sample is placed in a small glass tube and inserted in the centrifuge (sometimes called a shake-out machine). After being spun for a few minutes, the amount of water in a sample can be read easily (Figure 7–22). Most centrifuges are equipped with internal heaters that warm the sample to ease the separation of

Liquid Storage 215

Figure 7-18. A tank thief is used to obtain samples of liquids at any desired depth in a tank. (Courtesy of Midland College.)

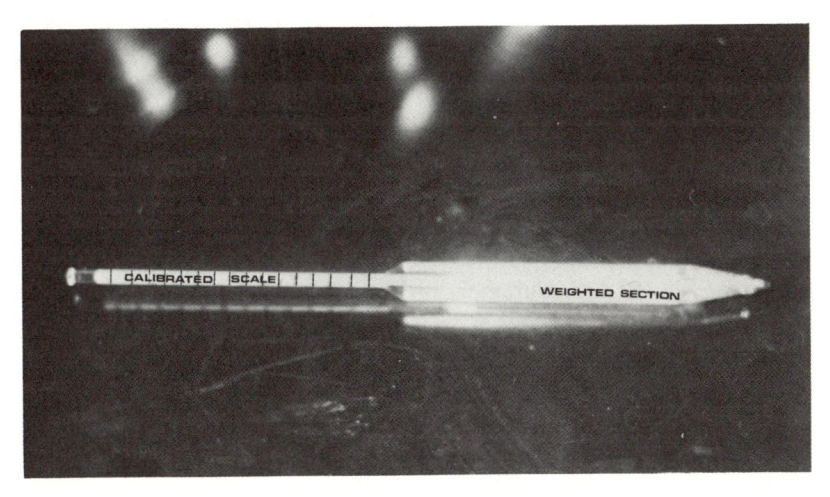

Figure 7-19. The weight in a hydrometer is selected so that it floats in liquid and indicates specific gravity or density on a calibrated scale. (Courtesy of Midland College.)

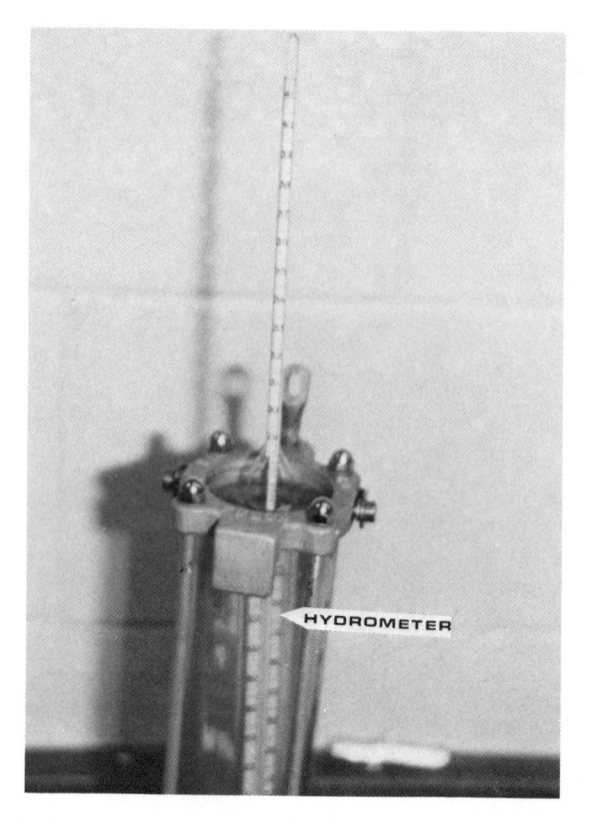

Figure 7-20. A hydrometer is weighted and calibrated to float in a liquid and indicate the fluid's density or specific gravity. (Courtesy of Midland College.)

oil and water. Sometimes a drop or two of demulsifier is added to a sample to aid separation. An alternate to the electrically driven centrifuge is a manual centrifuge (Figure 7–23) which accomplishes the same task but is operated manually.

Sampling and level measurement is done through an open thief hatch. The tank usually contains some gas as well as the liquid to be measured. The gas coming through the open hatch mixes with air and forms a combustible mixture. There is always a fire hazard above thief hatches, and appropriate safety precautions should be exercised.

Another hazard present above a thief hatch is lethal concentrations of gas. Hydrogen sulfide is present over the hatch of a tank containing sour oil. Since only a small concentration (20 parts per million) is lethal, breathing apparatus should be worn whenever working around thief hatches.

Liquid Storage 217

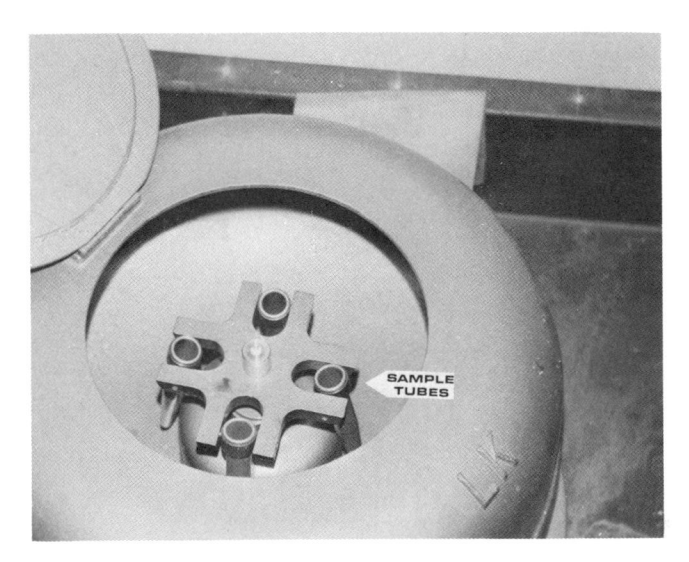

Figure 7-21. A centrifuge separates oil from water in a sample for an analysis of oil to be sold. (Courtesy of Midland College.)

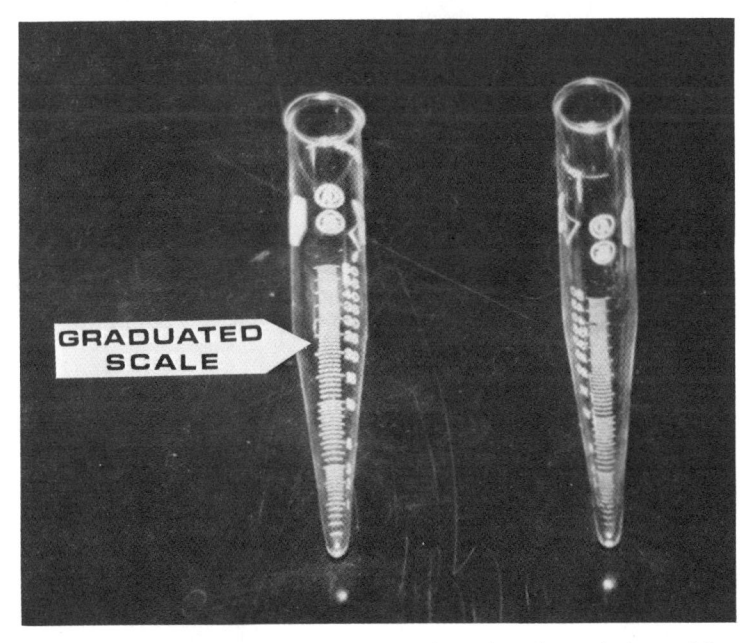

Figure 7-22. Centrifuge tubes have graduated scales to help read the tiny fractions of water sometimes measured. (Courtesy of Midland College.)

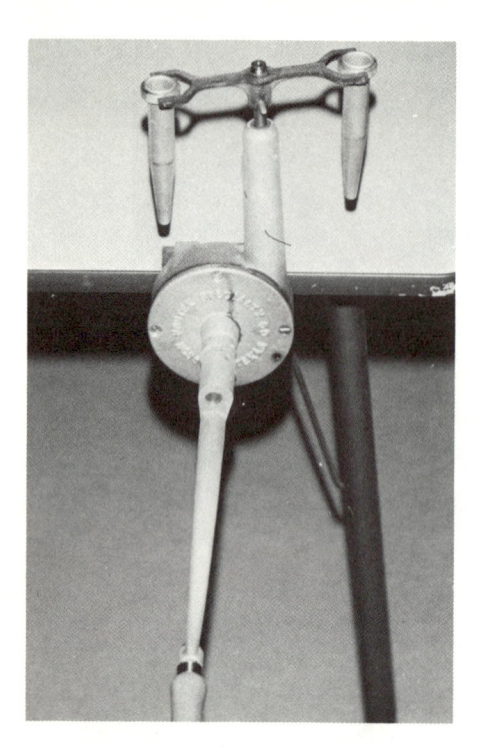

Figure 7-23. A manual centrifuge or shake-out machine performs the same function as an electric centrifuge. (Courtesy of Midland College.)

Liquid Pumps

As liquid is transferred from one vessel to another in a liquid processing facility, the pressure used to drive liquid is successively reduced at each stage. Eventually, the liquid pressure is reduced to the point that it cannot be pushed to another vessel. When this occurs, it is necessary to use pumps to move liquid.

There are two principal types of pumps used in petroleum processing. Positive displacement pumps are usually used when liquid is to be pumped at high pressure, while centrifugal pumps are used for low-pressure pumping. An exception is high-pressure centrifugal pumps used for high-volume, high-pressure pumping.

Positive Displacement Pumps

A positive displacement pump is diagrammed in Figure 7–24. It uses valves much like those in compressors. On each stroke of the plunger, a specific volume is moved. This volume is determined by cylinder diameter and piston stroke length. The prime mover for a positive displacement pump

Liquid Storage 219

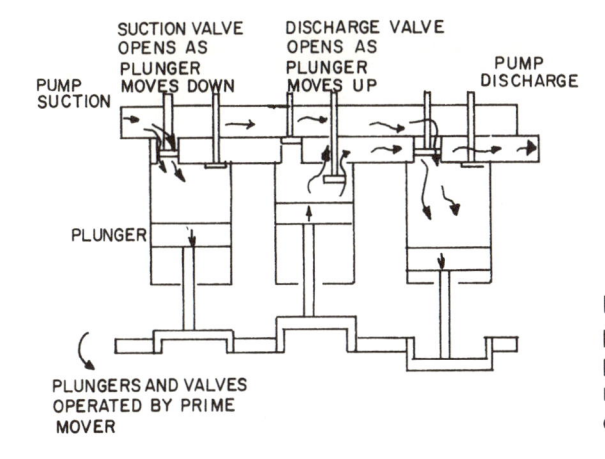

Figure 7-24. A positive-displacement pump uses plungers and valves to move incremental volumes of liquid.

Figure 7-25. One type of positive-displacement pump uses offset plungers under the discharge header.

is an electric motor, a gas engine, or a turbine engine. One application of a low-rate positive displacement pump is a chemical pump discussed in Chapter 6.

Several designs of positive displacement pumps are available. Figure 7–25 shows a pump using offset cylinders. Three-cylinder pumps are also available. Several cylinders are usually used to balance the crankshaft of the positive displacement pump.

220   Introduction to Petroleum Production

Figure 7-26. The plungers of a positive-displacement pump are operated by a prime mover to move liquid from the suction piping to the discharge piping.

On each suction stroke, liquid is pulled into a cylinder—sometimes rapidly enough to lower the suction pressure. On each discharge stroke, the pressure increases rapidly. Thus, the pressure on both the suction and discharge sides of a pump can vary. The pressure variation caused by a positive displacement pump may be great enough to interfere with the operation of the pump as well as equipment connected to it.

Pressure stabilizers or desurgers (Figure 7–27) are used to minimize the effect of pressure variation. As pressure begins to increase, the boot expands and absorbs the pressure surge. As the pressure begins to decrease, the compressed air forces the boot to contract—again absorbing the pressure variation. Pressure desurgers can absorb the pressure variations enough that the suction and discharge pressures of a positive displacement pump seem to be constant.

Positive displacement pumps operate at a constant rate determined by the shaft speed of the prime mover. Any effort to change this rate without changing the drive speed can result in excessive pressure and damage to the pump and other equipment. However, the overall rate of a pump can be controlled by using a by-pass line as shown in Figure 7–28. The rate downstream of the by-pass is regulated by the amount of liquid sent back to the pump suction.

Liquid Storage 221

Figure 7-27. Desurgers reduce the pressure variations resulting from individual plunger movement.

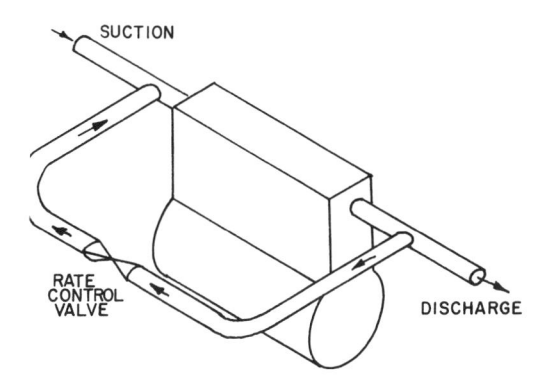

Figure 7-28. A by-pass valve may be used to control the discharge rate of a positive-displacement pump.

Centrifugal Pumps

A centrifugal pump moves liquid by rapidly rotating an impeller that slings liquid to the outside of a pump case as shown in Figure 7–29. Centrifugal pumps are generally used for lower pressures than positive displacement pumps because, as the discharge pressure increases, liquid slips past the impeller. Centrifugal pumps are available for application from very low rates of less than one gallon per minute to high rates of hundreds of gpm.

222 Introduction to Petroleum Production

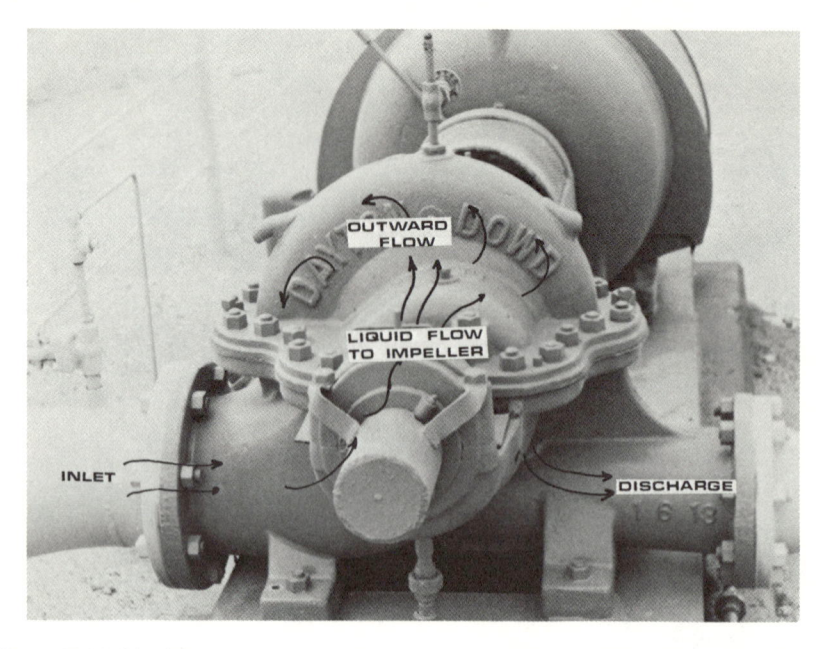

Figure 7-29. Liquid entering a centrifugal pump is slung outward by the impeller to the shaped pump body which directs liquid to discharge piping.

The impeller of a centrifugal pump (Figure 7–30) determines the rate and discharge pressure. The pressure and rate of an impeller are inversely related to each other—as discharge pressure increases, the rate decreases. There are usually several impellers that can be used for a given pump case. The front of a centrifugal pump may be removed and the impellar changed without difficulty.

Some large centrifugal pumps are driven with gas engines, but most small pumps of 100 gpm or less are powered by electric motors. Unlike position displacement pumps, the discharge rate of centrifugal pumps can be adjusted by placing a throttling valve in the discharge line as shown in Figure 7–31. As the discharge line is restricted, the rate drops as liquid slips past the impeller. Throttling the discharge of a centrifugal pump should not be done for long periods of time because the power delivered from the prime mover (usually a fixed value) and not used to move liquid is converted to heat in the pump. Excessive heat can develop rapidly and damage the pump or interfere with its operation. Any long-term rate change should be made by changing the impeller rather than by throttling.

Centrifugal pumps do not effectively "suck" in liquid. They require positive suction pressure to push liquid into the pump. If the suction pressure is too low for a given pump, the low pressure at the center of the impeller may

Figure 7-30. The shape of the impeller forces liquid flow to the pump case, while the impeller size determines flow rate and pressure.

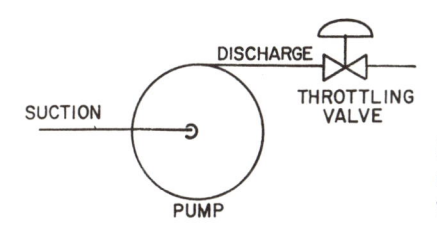

Figure 7-31. The rate of a centrifugal pump may be controlled for short periods of time with a throttling valve in the discharge line.

become low enough for the liquid to boil to vapor. Vapor in a pump causes a vapor-locked condition similar to gas lock in a subsurface pump. This cavitation in a pump can virtually destroy the impeller and pump body in very short order. Liquid erosion can also occur in the impeller and pump body unless these parts are made of hard metal. Corrosion of internal parts can also be a severe problem in centrifugal pumps. When pumps are to handle corrosive liquids, stainless steel or special alloys can be specified for the trim—the impeller, shaft, seals, and fittings. Solid particles, such as sand, can be pumped when suspended in liquid, but the pump case and impeller must be designed to resist the wear caused by the high-velocity particles hitting them.

High-Pressure Centrifugal Pumps

Figure 7–32 shows a high-pressure centrifugal pump. This pump is composed of a series of impellers that behave like the stages of an electric sub-

Figure 7-32. A high-pressure centrifugal pump utilizes a similar operating principle but has much heavier construction than conventional centrifugal pumps.

mersible pump. Each impeller adds some pressure to the liquid stream. By the time liquid passes from the suction to the discharge, the pressure can be raised to several thousand psig. Pump rates of several thousand gallons per minute are not uncommon.

Small-Lease Production Facilities

The equipment required to process fluids produced from a small lease are usually configured as simply as possible. The use of a centralized facility is usually dictated by economic considerations. It is often less expensive to install the equipment for a single facility to handle several wells than to install separate equipment for each well, provided the central facility is kept simple. This is not to say, however, that a central facility is designed for anything less than optimum effectiveness or minimum operating costs.

Figure 7–33 is an isometric sketch of a small-lease process facility. Incoming fluid passes through a header which can be used to divert production from one well through a test separator while processing all other fluid through a

Liquid Storage 225

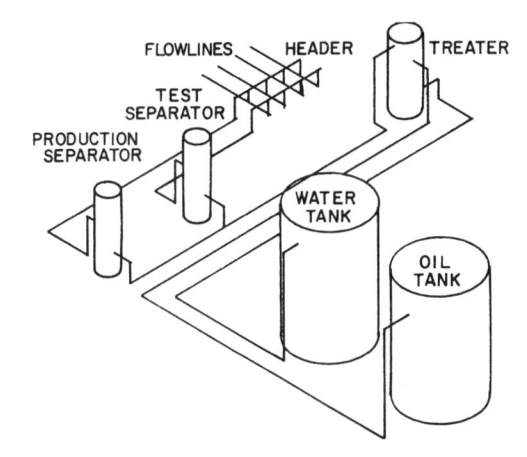

Figure 7-33. A small production facility must provide means of testing wells, separating gas, treating emulsions, and storing liquids.

production separator. The two separators remove gas from the liquid and provide pressure to push fluid on to the next process stage.

Fluid then passes to a vertical treater where oil and water are separated. Fuel for the treater and gas to operate pneumatic controls are taken from the natural gas stream and routed through a scrubber. Pneumatic supply gas is also run through a dehydrator.

Treated oil and water are routed to storage tanks where they are accumulated. This particular facility is arranged so that oil and water are hauled from the tanks to a refinery or pipeline by transport trucks. This is a usual practice for small, remote leases.

Figure 7-34 is a view of the process facility for a small gas-producing lease. In this lease, gas is produced from several zones and each zone must be served by independent process vessels. In this case it was best to put several process facilities at a single location. Incoming gas is routed to high-pressure separators where water and condensate are removed. The gas goes on to a dehydrator and then is compressed for shipment to a pipeline. Liquids are routed to a vertical, three-phase separator. Condensate is stored in one set of tanks, while water is stored in another. A vapor recovery unit is used to recover as much gas as possible from the volatile condensate.

These examples of production facilities are used for illustration only. They should not be taken as a pattern for lease facilities. There is no "best" configuration for all leases.

The facilities to recover and process fluids must be very carefully designed for each lease and for each well. On many leases, some wells may be produced by natural flow, while other wells require artificial lift. Even on leases

226 Introduction to Petroleum Production

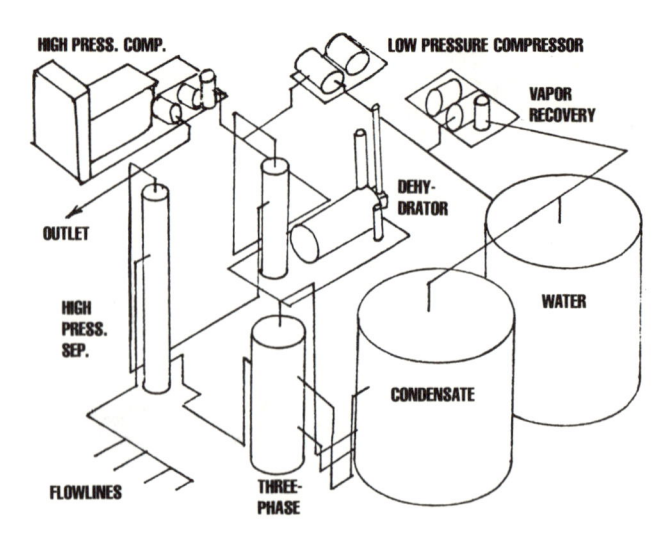

Figure 7-34. Even a small gas process facility can be a complex arrangement of separation, dehydration, storage, and compression equipment.

being produced by artificial lift, some wells may be driven by gas lift, while others are pumped with sucker rod systems, electric submersibles, or hydraulic pumps. The selection of artificial lift equipment must be made for each well individually.

The gathering system for a lease depends on the number of wells, their geographic spacing, the production rates, and the requirements for fluid separation. Some smaller leases do not use headers, but most modern installations use central facilities which do require headers. Larger leases or those producing from multiple zones may have remote headers as well. Gathering or distribution systems are also required when centralized hydraulic pumping systems or gas lift systems are used.

There is no definitive pattern for the process facilities of a small lease. The facility must be specifically designed for a particular application. A simple application may involve only two-phase separation, while the next facility may be a combination of separation, treating, dehydration, sweetening, storage, and transportation equipment. Separation and treating equipment must be selected based not only on initial and operating costs but also on fluid characteristics. The process facilities often must be designed to optimize installation cost against operating costs and maintaining API gravity and value of the petroleum products.

Large Production Facilities

Most small leases do not have enough oil and gas production to justify any but the simplest production facilities. These facilities are adequate for the production needs, but certain equipment and facilities are available which improve efficiency. These methods are sometimes costly and their use can be justified only when large volumes of hydrocarbons are processed.

Volume 3 of this series discusses some of the facilities that can process large-volume production streams. Other topics usually associated with larger leases are water disposal systems, water injection facilities, and the equipment used for gas injection and improved recovery techniques. Use of electrical equipment and systems are discussed, and some of the newest instrumentation and control equipment are covered.

Index

A

Absolute pressure, 159
Acid gas, 167
Acidizing, 4
Adjustable choke, 23-25
Air-balanced pumping unit, 56
Allowable (daily production rate), 22
American Petroleum Institute (API) 59,
    63-64, 68-70, 73, 78-79
American National Standards Institute
    (ANSI) pipe, 119
Anchors, 17-18, 83
Annealing, 69
Annulus, 15-17, 49-50
    fluid level, 85-87
Aquifer, 5
Artifical lift, 9, 45ff
Asbestos pipe, 120

B

Back-pressure valve, 128
Ball sealers, 4
Ball valves, 127
Beam pumping units, 56-59
Bellows orifice meter, 38
Block and bleed valve, 128
Bottom water drive, 3
Bottom hole chokes, 25
Bridge plug, 4
Burner controls (treaters), 183
Butress thread, 13
Bypass (compressor), 166

C

Carbon dioxide, 167
Casing
    intermediate, 5
    oil string, 5
    surface, 5
    vacuum pump, 111
Cement-lined pipe, 121
Centrifugal compressors.
    See Compressor(s).
Check valves, 88-89, 131
Chemical pumps, 88, 178
Chemical treating, 88, 178
Choke, 23
    adjustable, 23, 25
    bottom hole, 25
    gas lift, 46

long-nosed, 29
positive, 22-23
Circumferential displacement, 73
Coating, 210
Compression packer, 15-17
Compression
  multi-stage, 160
  ratio, 159
  stages, 98
Compressor(s), 51-52, 158ff
  bypass, 166
  centrifugal, 161
  controls, 165
  coolers, 165
  positive displacement, 158
  prime movers, 163
  skid-mounted, 166
  suction scrubbers, 165
  vapor recovery, 171
Condensate, 171
Coning, 22
Control, 21
  separator level, 141
  separator pressure, 141
Coolers, 165
Corrosion, 69, 88
  tank, 210
  coating, 210
Couplings
  pipe, 121
  sucker rods, 69-70

D

Data gathering. *See* Subsurface equipment.
Dehydrators, 152ff
  dessicant, 155
  glycol, 152
  instrument air, 156
Density, 4, 6
Dewatering, 45
Direct heaters, 27
Diverting valves, 128
Dynagraph, 86, 91-93
Dynamometer, 85-86

Dynamometer card, 87

E

EUE (externally upset tubing), 12-13
Electric
  motor, 61-62, 96-97, 163
  valve operators, 134
  submersible lift, 46
  submersible pumps, 96-97, 112-113
  submersible pumping system, 112-113
Electrostatic treating, 193
Elevators, 71-72
Emulsifying agents, 175
Emulsions, 175
  reverse, 176
Energy (reservoir), 1-2
Engine
  gas, 163
  hydraulic, 105-110
Enhanced recovery, 2
Evaporation, 172
Explosive fracturing, 4

F

Facilities (production), 224
Fiberglass pipe, 119
  high-pressure, 124
Fishing tool, 19
Flow
  behavior, 137
  control, 21
  natural, 9ff
  two-phase, 30
Flowing well, 10
Flowline, 26, 118
  high-pressure, 124
  installation, 123
  maintenance, 123
Fluid flow (gathering systems), 137
Fluid pound, 90-91
Fracturing
  explosive, 4

hydraulic, 4
Free-water knockout, 178
Friction, 21
   loss, 118

G

Gas
   compressors, 158ff
   condensate, 171
   controls (treaters), 185
   engine, 60, 163
   lock, 92-93
   meters, 33-38
   processing, 139ff, 172
   separation, 140
   sweetening, 167
   turbine engines, 163
Gas lift, 45-46, 111
   applications, 53-54
   continuous, 49
   intermittent, 49
   principles, 48-49
   valve, 48
Gate valve, 128
Gathering systems, 114
   axial, 115
   design, 138
   radial, 115
   trunkline, 115
Gauges. *See* Pressure.
Gear reducer, 57-58
Gearbox, 57, 63
Gels (blocking), 4
Glycol
   dehydrators, 152
   services, 156
Gradient, 7
Gun-barrel treating, 199

H

Head (hydrostatic), 7
Headers, 125
Heat transfer, 29-30
Heat treating, 177
Heaters, 26-29
Heating, 177

High-pressure flowlines, 124
Hold-downs, 76-77
Horizontal separators, 148
Horizontal treaters, 189
Hydraulic fracturing, 4
Hydraulic lift, 46
Hydraulic pumping, 105, 112-113
Hydrocarbons, 1
Hydrogen embrittlement, 68-69
Hydrogen sulfide, 68, 167
Hydrostatic head (pressure), 7, 45

I

Impeller (electric submersible pump), 98-99
Improved geometry pumping unit, 56
Indirect heaters, 27-29
Inferential meters, 38-39, 41-42
Inhibitor
   corrosion, 88
   scale, 88
Insert pump, 78
Installation (vertical treaters), 187
Instrument air dehydrators, 156
Intermediate casing, 5

J

Jet pump, 107-108

K

Knockout *See* Free-water knockout.

L

LPG, 61
Laminar flow, 21, 118
Level control
   separators, 148
   treaters, 189
Level gauge (glass), 143
Lift
   artificial, 9, 45
   natural, 9
Liquid meters, 39-42
Liquid processing, 175ff
Liquid storage, 202ff

Low-pressure flowlines, 119
Lubrication, 63, 96

M

Maintenance, 64-66, 96
  flowline, 123
  hydraulic pumping systems, 109-110
Manometer, 35, 37-38
Mercury meter (manometer), 35
Metallurgy, 11-12
  Meter, 33
  bellows, 38
  gas, 33-39
  inferential, 38-39, 41-42
  liquid, 41-42
  liquid orifice, 42
  mercury, 35, 37
  orifice, 35, 37
  positive displacement, 38-40
  two-phase, 42-43
Minimum yield, 11-12
Motors, 163

N

Natural flow (effect on artificial lift), 110
Natural gas,
  acid gas, 167
  sulfur recovery, 169
Natural lift, 9
  with gas lift, 46
Needle valve, 128
No-go, 19

O

Oil string casing, 5
Oil-water emulsions, 175
Open pits, 211
Operators, 131
  electric, 134
  manual, 131
  pneumatic, 133
Orifice meter, 33, 35, 38, 42

P

Packer, 4, 15-17
  unseating, 17
Paddle meter, 42
Paraffin, 4-5, 88, 118
Peak torque, 59
Permeability
  relative, 2, 4
Petroleum, 1
Pipe
  asbestos, 120
  cement-lined, 121
  couplings, 121
  fiberglass, 119
  high-pressure fiberglass, 124
  high-pressure steel, 119
  plastic, 120

Pits, 199, 211
Plastic pipe, 120
Plug valves, 127
Plunger lift, 46, 54
Pneumatic operators, 133
Polarity, 195
Polished rod, 66-67
Portable base, 62
Positive choke, 22-23

Positive displacement meter, 38-41
Power fluid, 107
Power fluid conditioning, 108-109
  central, 108
  solo, 108-109

Pressure, 26
  absolute, 159
  control (separators), 141
  flowing, 26
  gauges, 148
  regulator, 51
  relief, 143
  shut-in, 26

Primary recovery, 2
Prime movers, 61-62
Process (liquid) 175ff
Production equipment (subsurface), 10
Production facilities, 224

Production packer, 15-17
Proration, 54
Pumps
  casing vacuum, 111
  centrifugal, 221
  chemical, 178
  displacement, 78
  double-acting, 105
  electric submersible, 96-101
  high-pressure centrifugal, 223
  hydraulic, 105
  insert, 76-78
  liquid, 218
  positive displacement, 218
  single-acting, 105
  stationary barrel, 74-75
  three-tube, 78
  traveling barrel, 75-78

Pumping units, 56
  air-balanced, 56
  base, 63-64
  beam, 56
  derrick, 59-61
  improved geometry, 56
  loading, 61-63

R

Radial gathering systems, 115
Rate control, 21
Ratio (compression), 159
Recovery
  enhanced, 2
  primary, 2
  secondary, 2
  Regulator, 51
Relative permeability, 2
Reservoir
  energy, 1-2, 21
  pressure, 1-2, 45
Retention time, 148
Rod stretch, 83-85
Rod string, 66
  Round thread, 13
Rupture disk, 143

S

Safety, 196
Saltwater disposal, 2
Sand control, 5
Sand plug, 4
Saturation, 2, 4
Scale, 4, 88, 118
  inhibitor, 88
Schedule (ANSI), 119
Scrubber, 31
  suction, 165
Seal section. See Subsurface equipment.
Seating nipple, 19, 80
Secondary recovery, 2
Selective stimulation, 4

Separation
  condensate, 171
  electrostatic, 193
  gas, 140
  oil-water, 176ff
  two-phase, 140

Separators, 30-31, 140
  bottom-hole, 82
  gas, 99, 101
  horizontal, 148
  level control, 141
  other, 152
  pressure control, 141
  vertical, 140

Sheaves, 57
Shot peening, 69-70
Shroud (motor), 101
Sight glass, 143
Sizing (valves), 137
Skid mounting, 166
Slim hole couplings, 70
Slippage, 53
Slips, 15
Solids (wellbore), 95-96
Solution gas expansion, 3
Solvents, 88
Specific gravity, 6-7
Sprayed-metal couplings, 70
Standing valve, 19, 47, 75

Stationary barrel pump, 74
Steel pipe, 119
  high-pressure, 124
Storage
  open pit, 211
  liquid, 202ff
Stretch (sucker rods), 71
Stuffing box, 66-67
Submergence (pump), 94-95
Subsurface equipment
  analysis methods, 85-86
  data gathering, 85-86
  electric submersible pumps, 103
  gas lift, 46
  hydraulic pumps, 105
  sucker rod pumps, 66, 79
Subsurface pump, 74
Sucker rod, 66
  care, 71-74
  makeup, 73
  pump, 74
  stretch, 71
Sucker rod pumping, 46, 57, 83, 112
Suction scrubber bottles, 165
Sulfur recovery, 169
Surface casing, 5
Surface equipment
  electric submersible pumps, 103
  gas lift, 50
Surface tension, 4
Sweetening, 167

T

Tank level, 211
Tank sample, 211
Tanks, 202ff
  agitators, 207
  coating, 210
  corrosion protection, 210
  fiberglass, 210
  gun-barrel, 200
  pressure relief, 210
  risers, 206
  roll lines, 207
  siphons, 207
  sumps, 206

thief hatch, 207
vapor recovery, 171
walk and stair, 210
Tapping, 93
Temperature (display), 143
Tension packer, 15
Thermometers, 143
Thief hatch, 207
Thread, 13
Three-tube pump, 78
Torque, 57, 62-63
Traveling barrel pump, 75-76
Traveling valve, 75
Treaters, 180ff
  burner controls, 183
  electrostatic, 194
  gas controls, 185
  horizontal, 189
  level controls, 183
  safety, 196
  vertical, 181
Treating, 176ff
  chemical, 178
  electrostatic, 193
  gravity, 177
  gun-barrel tanks, 200
  open-pit, 199
Trunkline gathering systems, 115
Tubing, 11, 80
  anchor, 80
  dimensions, 15
  strengths, 15
Turbine engines, 163
Turbine meters
  gas, 39
  liquid, 41
Turbulent flow, 21, 118
Two-phase metering, 42-43
Two-phase separation, 140

U

Undertravel, 85
Unseating, 17
Upset
  external, 12-13
  internal, 12-13

needle, 128
   plug, 127
   pneumatic operators, 133
   relief, 143
   sizing, 137
   standing, 17, 74
   traveling, 74
   vent, 143
   wafer, 128
Vapor recovery, 171
Vapor recovery unit, 172
Vent valves, 143
Vertical separators, 140
Vertical treaters, 181
Viscosity, 4
Vortex meter, 42

W

Wafer valves, 128
Water bath, 27
Water injection, 2
Water removal, 152
Waterflooding, 2
Wellhead, 5
   heaters, 27
   sucker rod pumping system, 87
Wetting tendencies, 2

Y

Yield, 11